I0481127

Engineer, Set Refueling Power

GR Dornfeld

Engineer, Set Refueling Power

BIOGRAPHY

GR Dornfeld earned a B.S. Degree in Aeronautical Engineering from Boston University and a commission in the Air Force Reserve through Air Force ROTC in 1960. He was called to active duty in January 1961, completed Undergraduate Navigator School at James Connolly Air Force Base, Waco, Texas, and Strategic Air Command (SAC) Combat Crew Training in the KC-97 Air Refueling Tanker at Randolph Air Force Base, San Antonio, Texas in 1962. He served on a crew in the 19th Air Refueling Squadron, in SAC, at Otis Air Force Base, Massachusetts, in KC-97 Tankers until late 1965.

Upon release from active duty, he worked for Hamilton Standard as a Propeller Installation Engineer in Windsor Locks, Connecticut, while flying for the Air Force Reserve as a navigator in the C-124 aircraft. He left Hamilton Standard to join GE Large Jet Engine Division in Cincinnati, Ohio in late 1969, as a CF6 Installation Engineer. During that time, he flew in the Ohio Air National Guard in their KC-97L Tanker until 1973.

The author transferred to the GE Small Jet Engine Division in Lynn, Massachusetts and flew as a navigator in the Air Force Reserve in the C-123K Transport for a while. He completed 31 years with GE, attaining the title of senior staff engineer in external configurations design by the time he retired in 2001. He also attained the rank of lieutenant colonel in the Air Force Reserve and an aeronautical rating of senior navigator, by the time he retired from the Air Force. GR Dornfeld is married to Marilyn and has three children, Michael, Scott and Paula, all married, and five grandchildren, Matthew, Chelsea, Rina, Shir and Maayan.

Preface

This book was written in order to capture my experiences during the years I flew as a navigator for the active duty Air Force, the Ohio Air National Guard, and the Air Force Reserve, and throughout my career in engineering. I was not a hero, nor do I claim to be, and I did not serve tours in Vietnam as so many of my peers have. However, I flew during the cold war years of 1961 through 1975, and was subject to call-up at any time. I have spent long hours thinking about my experiences, but, previously, have not been able to put it all together. My wife, Marilyn takes the credit for encouraging me to write this book. I would not change a day of my life as a rated Air Force navigator. The book may include too many photographs, but I felt they were necessary in order to give the reader a feeling of what I experienced. In the interest of privacy, the names of most people in the book have been changed.

Engineer, Set Refueling Power

Engineer, Set Refueling Power

Table of Contents

Engineer, Set Refueling Power

Engineer, Set Refueling Power

CHAPTER 1: GROWING UP
Trip to the Kosher Slaughterhouse

Mother tells the story of my trip to the kosher chicken slaughterhouse when I was three years old. I was playing in the yard when I apparently had a yen to see the chickens. I hopped on my tricycle and pedaled my way to the slaughterhouse, perhaps one half mile away. When Mother discovered me missing, she panicked and called my father at work. He borrowed his boss's car and my parents went up one street and down the next looking for me. They finally spotted me as I waited for the traffic to clear, so I could cross North Common Street, in Lynn, Massachusetts. I had already crossed South Common Street and the Commons itself. I waited, with my chin on the handlebars of my tricycle, for the traffic to allow me to cross this busy thoroughfare.

Parents' Wedding Photograph
Photo by Author

Engineer, Set Refueling Power

Ear infection
When I was four years old, I experienced multiple occurrences of an ear infection. In those days, the standard remedy was to remove the tonsils. Also, in those days, the doctor made house calls. I remember Doctor Shulman laying me down on our wooden kitchen table, and putting the ether mask over my nose and mouth. His words were, "Be brave, like the soldiers in the army." He then proceeded to remove my tonsils. Of course, my parents played the cruel trick of promising me I could eat all the ice cream I wanted after the operation. It was 1943 and the war was raging in Europe and the Pacific.

In 1945, the war was finally over, and there was to be a monstrous parade in Lynn to celebrate the end of the war. I was pumped up. I heard an unfamiliar airplane engine, at low altitude. When the vehicle came over the house, I saw it was a blimp. Wow, what a sight. I must have been really wound up, because Mother told me I could not go to the parade. I really must have misbehaved for her to say that. It was then I had my first feeling of *deja vu*. I just felt I had been there before. This feeling overcame me a couple of times in my youth, but never again when I grew up.

My vague memories of my father at an early age
I have little memory of my father when I was growing up. One I recall is when he drove a truck for a living, delivering mattresses locally. It was the end of the workday, and Mother walked with me in the Lynn Commons. She spotted my father, as he was about to return the truck to the business parking lot. I remember him hoisting me up into the passenger seat of the truck, and driving me back with him to the parking lot for the business. It was fun! Mother told me my father always had a job during the Depression. He always hustled, and put out more than was required. When he drove a truck, and returned it to the "barn" at the end of the day, he washed it, even though it wasn't a job requirement. I was proud of him for doing that. He always seemed to get on the good side of his boss, which is probably why he always had a job.

Engineer, Set Refueling Power

My father had a critical defense job during World War II, and consequently he was not drafted. It also helped he fathered a child by that time. He worked at the General Electric River Works (the same plant where I worked for 28 years), in a tool crib on the manufacturing floor. Right after the war, he left GE and went back to driving a truck. Many years later, I was at a party, when I was introduced to an elderly man who was my father's boss at GE. I so wanted to pump this man for information about my father, but the time and place were not appropriate, and I let it go. Not much later, I heard this man was admitted to a nursing home for Alzheimer's disease. This was a significant missed opportunity.

Daily, as I awoke, and while my father got ready for work, I remember the sound of a steam-powered train engine, which was trying to gain traction from the station in Central Square, Lynn. It obviously had a heavy load, as it made noise, as follows: chug .chug .chug, chugchugchug chug. silence........chug chug chug.chugchugchug. Apparently the load was so heavy, and the friction between the driven steel wheels and the steel track was only so much. In later years, thinking about this, I came to the conclusion the train was loaded with war material from GE!

My Double Promotion from First to the Second Grade

I was enrolled in Cobbet Elementary School when I was six years old. This was right after World War II, and the public schools were overcrowded. After two weeks in first grade, I was moved to the second grade. Wow, a skip promotion. Apparently, everyone who attended private kindergarten (as I did) was selected to skip the first grade, to alleviate the overcrowding. I hadn't mastered addition, and when I moved to the second grade, my teacher was instructing subtraction. I was seated in the rear right of the classroom, next to Charlie Rosen. The teacher, Miss Hale, said to Charlie, "Show Gerry Dornfeld what we are doing." Charlie whispered to me, "I don't know what we are doing, either." So I started six years of elementary schooling completely over my head. I finally started catching on to the coursework in the seventh grade. By the way, I have something to say about Charlie's father, later on.

Lesson in Truth Telling

I was in the fifth grade at Cobbet Elementary School, in Lynn, Massachusetts. The teacher was Miss Hussey, who was also the school principal. The Ringling Brothers Circus came to The Boston Garden. Mother and her girlfriend, Ruth, decided to take me and Ruth's Arnold out of school for the day to see the circus. There were quite a few other absences that day in school.

The next day all the other students came to class with notes from their parents, requesting their absence be excused, due to illness. Ruth and Mother sent us to class with notes requesting an excused absence because they took us to the circus. Miss Hussy let all the little liars off the hook, but punished Arnold and me because our mothers told the truth as to why we were absent. Arnold and I had to make up the hours lost that day by staying after school the rest of the week. This definitely didn't make me want to emulate George Washington for admitting he chopped down the cherry tree.

More Memoirs from School

I was in the sixth grade in Cobbet Elementary School, and we were studying Economics. I was still completely lost, since the second grade. The teacher told me if I didn't do my assignment, I would be held back a year. The assignment was to assemble a scrapbook of pictures from magazines of items related to the subject of Economics. Boy was this a wakeup call. I went home and cut out pictures from magazines of anything remotely affecting Economics, and pasted the scraps onto a page. I passed. This teacher did me a favor, and made me wake up.

In the seventh grade, I discovered I enjoyed, and was good at, the subjects of algebra and science. I had an uncle and a great uncle who were civil engineers, and I started thinking along the lines of an engineering career. As my grades came up, Mother made an appointment to see the junior high school principal, Mr. Dald. Mother was worried we could not afford college. Mr. Dald told her not to be concerned, but for me to concentrate on getting the highest grades possible, and the money would come.

He was right, and I really appreciated his input. It was 1951 and I read Aviation Week magazine every week. The Korean War was raging, and there were many articles on advancements in aviation, especially articles related to the war. I decided to become an aeronautical engineer right on the spot, and never wavered.

Years later, Mother told me about the obituary on Mr. Dald in the Lynn newspaper. The thought had occurred to me to go back and visit with him and tell him how his advice was spot on, and I was successful, in part due to his confidence and advice. I wanted to tell him I earned a college degree (B.S. in Aeronautical Engineering), a commission in the Air Force, and an Aeronautical Rating of Aircraft Navigator. Also, I was serving as a crewmember in KC-97 Tankers at Otis Air Force Base. Alas, I procrastinated too long, and the opportunity was lost forever.

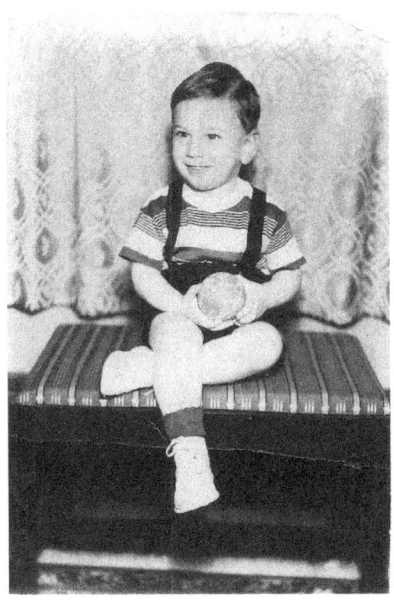

Formal Portrait of Author, Age 3; Circa 1942
Photo by Author

Author and Sister, Sandi, taken about 1948
Photo by Author

Morris and Rose Glass

We lived in a 36 unit apartment building. There were three floors of 12 units each. Our apartment was the furthest from the street, on the right side, first floor. Above us lived siblings Kate, Sharon and their brother Sam, and they were all single. Morris and Rose Glass lived above them, on the third floor. Morris delivered milk, door to door. I remember his horse-drawn wagon, with large blocks of ice piled on top of the milk products to keep them from spoiling. We youngsters, on a hot summer day, grabbed a chunk of ice and sucked it until it completely melted in our mouths. Of course, there were remnants of horse manure mixed in with the ice, but that didn't bother us. We were young and were going to live forever.

Morris and Rose had three sons, all older than me. I only remember John, their youngest. One of their older boys, who I only vaguely remember, married a gorgeous redhead. Within a year, they were divorced. I think fidelity on her part was the issue. I think (but I am not sure) I came into the possession of a baseball signed by the 1948 Boston Red Sox through a connection with the Glass family. My memory tells me they were related to a famous first baseman for the Boston Red Sox. Anyway, I still have the baseball signed by the whole team. I am sure the baseball is worth a few dollars.

My First Job in a Supermarket
Susan Fried lived in our apartment building and worked part-time as a bookkeeper for the North Shore Market in Lynn. She recommended me for a part-time grocery clerk position when I was 13 years old. I started work at the wage of $.49 per hour. The year was 1952. I was a year too young to legally work, but I was blissfully unaware of this fact. One day, the city inspector paid the North Shore Market an unannounced visit to check all the part-time workers were at least 14 years old, as the law required. I was shocked at the time and left work immediately, figuring I was in deep trouble. The owner tried to convince me to stay at work, but I didn't. I now know the city worker was just looking for his bribe, and probably got it.

I had responsibility for stocking the shelves with groceries on Friday afternoon after school, and Saturdays. The business was in an old building, and the groceries were stored in the basement. As I was about to go downstairs, the light switch for the basement was on the right hand side next to the highest step. The basement stank of ammonia, and the noise was troublesome, from the refrigeration units. And there were rats. I was fearful someone would shut the lights off without knowing I was in the basement. It was dark without the lights, and I would not be able to find my way back up the stairs to get to the light switch. Luckily, it never happened.

The cases of wooden matches were stacked on the highest shelf on the main floor storage unit. I was in the habit of climbing to the first shelf, reaching up for the case of wooden matches, and then dropping the case to the floor. This always worked until one day, when the case hit the floor, it lit off and a nice fire erupted. It was put out quickly, but the stink of sulfur permeated the store all night. It was bad and I was embarrassed.

I heard the Stop & Shop Supermarket on Union Street was looking for a part-time clerk for their produce department. I walked to Union Street and applied for the job. The hiring manager was Mr. Rosen, the father of Charlie Rosen who sat next to me in second grade. Mr. Rosen told me I was hired and I was to report on Friday. I asked him if I could start the following week, as this week was the Jewish High Holidays (Mr. Rosen was also Jewish.) He told me abruptly, "No, It's either Friday, or no job." I passed on the job and was upset a fellow Jew would treat me this way.

I secured a part-time job at the new Stop & Shop in Vinnen Square, Salem, Massachusetts. Apparently, my Aunt Ginger played cards with Mrs. Mild, the wife of the Store Manager Morris Mild. I started work two weeks before the store opened, by loading the shelves with the first store-load of groceries. I reported directly to Lewis Danniel, the Grocery Manager. Of all the part-time employees, I was the only one without access to a car. When the dismissal bell rang, I left school precisely at 2:15 P.M. and ran to catch a bus departing a few blocks from school at 2:40 P.M. to Vinnen Square. Occasionally, I bummed a ride to work from a peer whose first name was Justin. He had access to his mother's car, a gorgeous baby blue 1953 Ford Convertible. Justin had life too easy, and treated the car with little respect. At each stoplight, he peeled rubber as he popped the clutch and accelerated as fast as he could. I loved that car.

The other part-timers had access to either their own car or their parent's car, and they didn't need the job. I needed the job. I was hired at $.89 per hour. The other clerks had a bad habit of reporting late for work (as I said, they didn't need the job.) The grocery manager was upset with the tardiness and decided to make an example of the next part-timer who was late.

1953 Ford Convertible
Photo by Author

I left school exactly on time one day, and caught the bus right on schedule. However, the driver was on his maiden run as a new employee, made a wrong turn, and drove down the beach road instead of the one parallel to it. All the riders yelled at the driver, "You are going the wrong way." He regained the correct route a block away, but I was five minutes late. I tried to explain to the grocery manager why I was late, but he didn't believe me, and I was docked one hour pay. I was furious, as I needed the money, and it was not my fault. Life wasn't fair!

I worked for Stop & Shop Supermarket right through college, part time. In my junior year in college, I cut down on my work hours so I worked only on weekends. Management decided to expand the store hours on Saturday by three hours, to 9:00 P.M., and I was offered the opportunity to work 9:00 A.M. to 9:00 P.M. I was happy to work the 11 hours (with a half hour for lunch and supper) and quickly agreed to the hours. After two weeks, my hours were trimmed back to 12:00 P.M. to 9:00 P.M. Now I worked only eight hours (with an hour off for supper) and the other part-timers worked 9:00 A.M. to 6:00 P.M. I felt I was discriminated against, as I worked the same amount of hours as the others, on Saturday, but I lost any social life I might have had, by working <u>every</u> Saturday night. When I complained to the grocery manager, he said, "Tough. If you don't like it, then quit." I never forgave him for that, but I needed the job and I sucked it up. Again, life was not fair!

I have always been interested in aviation. I was born in 1939, so I was too young to be involved in World War II. As my teen-age years approached, I was fascinated with the aircraft from this era. The Massachusetts Air National Guard was based at Logan Airport in Boston in the late 1940s, and was equipped with the P-51 Mustang fighter aircraft (later redesignated as the F-51). As a youth, I attended summer day camp and daily was bussed to Lynn Beach for an afternoon of supervised play. There apparently was a big military exercise going on, one day, because there were many fighter aircraft, both P-51 and P-47 Thunderbolts flying at high speed over the beach, just offshore, and on the deck (very low level). The pleasing sound of the engine exhaust and propeller from the P-51 Rolls Royce Merlin piston engine/propeller remains with me to this day.

Author, 10, and his sister Sandi, three, June 1949
Photo by Author

My Interest in Engines

I was intensely interested in aviation, as well as automobiles, anything mechanical. A common denominator between aircraft and automobiles was the piston engine. I wasn't sure which way my career interest would lead, but I was certain it would involve either automobiles or aircraft, and specifically engines. On the aviation side, the cold war was starting, and there were many aircraft manufacturers involved in military aircraft design activity, using both piston engines and the early vintage jet engine. I was a sponge absorbing everything.

On the automotive side, there were a host of manufacturers, no longer in business. There were competing designs uniquely different from one another. I could differentiate between a Buick, Oldsmobile, Pontiac, Chevrolet, Cadillac, Ford, Mercury, Lincoln, Chrysler, Plymouth, Dodge, De Soto, Studebaker, Packard, and Rambler (have I left out anyone?) a city block away, because the designs were so different from each another. Of course, I grew up poor, so my family did not own an automobile during my youth. But I dreamed anyway.

When the new cars were unveiled, I walked to the nearest car dealer and looked at the new designs. I remember walking to the Lynnway to the Packard dealer in the early 1950s, to see their new design; the salesman knew I wasn't a customer, and kicked me out. My ultimate revenge is Packard is out of business, and the building is now a mattress retailer (Sleepys). The car repair bays are still attached to the building.

I completed the seventh grade, at Cobbet Junior High School, in Lynn, when the Korean War broke out on June 25, 1950. I followed the news on this war intensely, especially as regards to aviation. My aunt rented an apartment above another tenant who was a private pilot, and was building his own airplane. This tenant subscribed to Aviation Week Magazine, and gave the week-old issues to my aunt to give to me. I read those issues from cover to cover. I especially remember reading about the twin boom C-119 aircraft airdrops during the Chosin Reservoir Operation. Unlike my peers, who didn't really know what career field they were interested in, I decided to become an aeronautical engineer right there and then, and never wavered. I read every book in the Lynn Public Library on the subject of aviation. I specifically remember reading about the details of the piston engine carburetor from one book.

There was a newspaper notice announcing a fly-over of a jet-powered F-86 fighter at a certain time and date. I got a note from Mother to release me from class so I could go to the Lynn Commons to watch the flyby. I cannot recall actually seeing the flyby, but retired General Dwight Eisenhower came to make a campaign stop and speech during his run for the presidency, and I saw him, from afar.

American Volunteer Group (Flying Tigers)

On Sunday mornings there was a television documentary on the Pre-WWII Flying Tigers. This was the American Volunteer Group flying worn out P-40 fighter aircraft in China, under contract to the Chinese government, and fighting the A6M Japanese Zero aircraft, before the U.S. came into the war. Those were the days of the "rabbit ear" television antenna, and the signal was weak. The picture was mostly heavy "snow," but I was so fascinated with the aircraft I watched the entire program, week after week, even though most of the time I couldn't see anything.

American Aircraft in the Korean War

The Korean War was the military transition point from the piston engine-powered aircraft to the jet engine-powered aircraft. The Korean War started with leftover WWII era aircraft, such as the North American F-51 Mustang fighter, the Boeing B-29 Superfortress bomber, and a variety of piston engine-powered cargo aircraft. The Air Force quickly transitioned to jet-powered equipment, as it became available, such as the North American F-86 Sabre, the Republic F-84 Thunderjet, and the Lockheed F-80 Shooting Star, in order to fly combat with the Russian jet-powered MIG-15s.

The television evening national news consisted of a 15-minute segment hosted by John Cameron Swayze, called the "Camel (as in Camel cigarette) News Caravan," on the NBC news network. He frequently had gun camera combat footage of the F-86 and Mig-15 jet-powered fighters mixing it up in dogfights. The day of the jet powered bomber was not quite upon us at that point, but was in development, and the war was fought with the B-29 and B-50 bombers powered by piston engines. I absorbed everything written about the various aircraft and the principles of their operation.

Engineer, Set Refueling Power

General Electric Developed the First American Jet Engine

The General Electric plant in Lynn, Massachusetts was involved in the design and manufacture of jet engines, since 1942. This effort resulted from the GE work on turbo-superchargers before and during WWII. England actually started work on the jet engine, on the Allied side, with a design by Sir Frank Whittle during WWII. However, because England was involved in a bitter struggle for its very survival, it could not expend the effort to bring the engine design to fruition. GE, Lynn was selected in 1942 to take the Whittle design and develop it to maturity, which it did.

I remember the ear splitting sound coming from the GE test cell where they tested the engine (we didn't actually know where the loud noise was coming from, at the time. But, way later, I concluded it was from the test cell used to test the first design jet engine.) There was no sound suppression back then, and the engine had an open tailpipe for its exhaust. The neighbors might have complained, but the war took priority. We lived about a mile away, and the sound was still loud. Years later, I spent 28 of my 31 years at GE working at the Lynn plant. The small test cell for the first jet engine still stands, although the cell is no longer active. It is used as a historical monument to the jet engine development.

Ground Observer Corps

In the early 1950s, the Air Force was concerned the Soviet Union could fly bombers over the U.S. under the radar. The government set up a Ground Observer Corps to spot and report low flying aircraft. There was a unit at Beverly Airport, in a control tower. I took the bus to talk to the person in charge, in a raging snowstorm. I remember to this day the call sign of that Ground Observer Corps station; Echo Nectar Zero Two Black (EN02B). On the ceiling of the control tower was a compass rose. When an aircraft was spotted, the observer telephoned his higher in command and reported the aircraft compass bearing, approximate altitude and speed. Unfortunately, I was just under the age limit and they would not make an exception for me. Furthermore, when I was 14 years old, I either lost interest or the program was discontinued, as I do not remember following up.

RB-47 Jet Reconnaissance Bombers Testing the Borders of the USSR and China

During those uncertain years of the mid-1950s, the newspapers reported an occasional downing of a jet-powered B-47 bomber as it flew "near" the Soviet Union or Communist China border. There were unconfirmed reports of B-47 bombers penetrating the Soviet Union, also. To this day, there remains some controversy about this issue, but a search of the internet details an account of an actual penetration of the Soviet Union to obtain photographs of Russian airfields by an RB-47 aircraft (the "R" stands for the reconnaissance version of the bomber aircraft).

My Experience with Civil Air Patrol

At age 14, in 1953, I joined the Civil Air Patrol (CAP) Squadron, in Salem, Massachusetts. I had my first airplane ride in an L-19 light aircraft owned by the Tewksbury CAP Squadron at Tew-Mac Airport. I attended a CAP Summer Camp at Sampson Air Force Base (AFB) in New York for two weeks. This was an Air Force basic training base for new enlistees. All I remember from this experience was hearing a lot of chewing out and going through the tear gas chamber. The latter was not a fun experience. I went through the entire aviation education program sponsored by CAP, at the Salem Squadron, and passed the proficiency exam. This allowed me to go through Air Force basic training with one stripe, Airman 1st Class, which was a big advantage. However, I never needed it, as I had bigger goals in life.

The Movie, Strategic Air Command

In 1957, when I was in high school, the movie "Strategic Air Command" was released. The plot was about a major league baseball pitcher who was recalled to active duty in the Air Force in his former specialty as a WWII bomber pilot. James Stewart, who was an actual decorated Air Force bomber pilot in World War II, played the lead. His co-star was June Allyson. The movie started with Stewart pitching a professional baseball game and a mammoth piston engine-powered B-36 bomber taking off just above the ball field.

The movie follows Stewart as he is recalled to active duty, and checks out in the B-36 bomber. He later flew a mission in the arctic area, survived a crash landing, and transitioned into the new jet-powered B-47 bomber. The movie was somewhat schmaltzy, but the effect on me was electrifying. There was a scene in the movie where the new B-47 bomber wing was deployed from MacDill AFB in Florida, and a refueling with a KC-97 tanker was shown. Little did I know, later on, I would be intimately involved with that tanker aircraft. I was sold. I knew I was going to be associated with the Air Force in some capacity later on. I must have seen the movie a dozen times over the years.

Cessna L-19 Bird Dog. The author's first airplane ride was in an L-19 Bird Dog, owned by the Tewksbury, Massachusetts Squadron of Civil Air Patrol.
Photo Ref: http://en.wikipedia.org/wiki/L-19_Bird_Dog

My Aspirations to Join the Air Force

I wore eyeglasses since I was in high school. It was obvious I could never qualify for training as an Air Force pilot. I was also interested in science and math, and intended to become an aeronautical engineer. My career interests could have been satisfied as an engineer in the Air Force, or maybe a meteorologist. The Air Force had a program (for their meteorologists) where they put selected officers through an intensive mathematics-driven course of study at Massachusetts Institute of Technology (MIT), leading to nearly all the credits necessary to earn a M.S. degree in Mathematics.

After high school, I attended Boston University Engineering School, and participated in Air Force Reserve Officer Training Corps (AFROTC). Any student could join AFROTC in the basic course, consisting of the freshman and sophomore years. However, I needed to be selected to move on to the advanced corps of AFROTC, for the junior and senior years. In order to be selected, I needed to pass the Air Force Officer Qualification Test (AFOQT), pass the physical examination, and be "slated" for an Air Force specialty forecasted to be in need.

Qualifying for Navigator Training

Near the end of my sophomore year, in the spring of 1958, I took the AFOQT and physical examination at the Boston Army Base in South Boston. My eyesight was under 20/50 correctible to 20/20 with glasses; this was not good enough for pilot training, but was acceptable for navigator training. I was pleasantly surprised, as I was not aware I could qualify for navigator training. I met the selection board for the advanced corps of AFROTC and was selected. I was exhilarated.

In my junior and senior years at Boston University, AFROTC consumed all of the elective credits I was allowed, in order to earn my engineering degree. I had some interesting active duty officer instructors while in the program. I have changed their names to protect their privacy.

My First and Second Car

Every man remembers the history of the cars he owned throughout his life. I was crazy about cars while I was growing up, but my family could not afford one. I actually bought the first car in my family when I was a freshman in college. In 1957, I managed to save $175 and bought a well-used 1949 Pontiac from Sea Crest Used Car lot in Lynn, Massachusetts.

I was under age, at 17, to own a car, so Mother walked with me to the car lot on the Lynnway, after work, and she purchased it in her name.

Mother was nervous about letting me drive the car without her, but I paid for the car, and I had to take my friends out for a ride. The car was painted medium green and had a straight-eight engine with a three speed manual transmission. Of course, the floor was rotted out, as this was a New England car. The floor mats shielded me from much of the road spray when it was wet. The clutch chattered when let out and there was a low moaning sound coming from the rear of the car.

Since the Pontiac had snow tires, I assumed the moan was coming from the aggressive tire tread pattern. Also, when the engine was hot, and the car was in heavy traffic, the engine suddenly quit. Although I didn't know it at the time, the problem was vapor lock, a common fuel system problem with cars of the period.

I sold the Pontiac two years later for $150, a loss of $25 from the original purchase price. I had to have a Ford or Mercury with its melodious flathead V-8 engine. I searched the daily newspaper classified ads for a car meeting my needs. I found a 1950 Ford Coupe for sale in Swampscott, Massachusetts. After calling on the ad, the owner told me the car was not registered, so a test drive was not possible. But I could test the clutch in his driveway. He wanted $350 for the car and was not flexible on the price. I could not afford it, nor did I look at it.

I found a 1950 Mercury Sport Sedan in a used car lot on the Marsh Road, in Saugus. After taking a bus to look at the car, the salesman could not get the engine to fire, although it cranked. He told me the car had an electric fuel pump and it was why it would not fire. I did not understand the rationale for the electric fuel pump impacting the ability of the engine to start, so I did not buy the car.

Obey Motor Sales on Summer Street, in Lynn, had a 1950 Mercury Coupe on the lot. I fell in love immediately, in spite of the $295 price, and the grinding noise the transmission made when shifting into 2nd gear. Also, the radio didn't work. I purchased the car for full price, put down the $150 from the sale of the Pontiac, and paid out the balance over the summer from my job at the Stop & Shop Supermarket.

Pontiac Straight Eight Engine
Ref: http://en.wikipedia.org/wiki/Pontiac_Straight-8_engine

While still within the 30 day 50/50 warranty, the emergency brake cable snapped. Obey Motors sent the car across the street to a Gulf Service Station, and the mechanic fixed the cable. I then brought the car back for the grinding noise when shifting into 2nd gear. While the used car manager chain-smoked cigarettes, the mechanic told him the clutch was *shot,* and had no more adjustment left. The manager said, "There had to be (an adjustment), there had to be." A 50/50 warranty meant the dealer jacked up the cost of the repair by 100% and took off 50% for his contribution. I knew I was screwed!

I was paid $50 every quarter for participating in the Advanced Corps of AFROTC, and took one payment and the car to McManus Garage on Western Ave., in Lynn, to get rid of the transmission grinding between gears. The mechanic told me I needed a clutch, which was just about $50. I had it repaired, but the grinding noise was still there. When I brought the car back, he told me the problem had to be with the transmission. I had no more money to address this problem, so I lived with it.

1950 Mercury Coupe. This was actually the second 1950 Mercury Coupe I owned (I purchased it later in life; see a later chapter). Photo by Author

Screwed by a Service Station in Revere, Massachusetts

I loved the Mercury in spite of its mechanical deficiencies. The body was in terrific shape for a New England car. I drove the car from Lynn to Revere Beach Metropolitan Transit Authority (MTA) station each day to commute to Boston University. One day, upon returning from school, the car would not start. I got a push the two blocks to a service station, and left it to be repaired. Later, after I paid to have the fuel pump replaced with a rebuilt, and drove off, the engine made a "whump...whump" sound. I now know the noise came from the fuel pump, but I foolishly sucked it up and drove off. While driving to work at the Stop & Shop Supermarket, about half way there, the engine quit again.

This time it was caused by a failed ignition rotor (it was broken!) I had the rotor replaced and the second rotor broke a short time later. This definitely was not fun. I found out somehow a tiny clip probably worth a penny was missing. This part prevented the rotor from orbiting as it rotated. When it orbited, it contacted the distributor cap and broke. The service shop not only did the lousy job on my rebuilt fuel pump, but probably cannibalized the clip when the car was in their shop. I sold the car after I married Marilyn and was about to leave for active duty with the Air Force. I believe I got $125 for the car. What I wouldn't do to have this car today.

Six Turning, 2 Burning

Two of my engineering school classmates were Air Force veterans. One had been a copilot in the B-47 aircraft, and the other a navigator in the B-36 bomber, both in SAC.

I was most impressed with the stories of the navigator on the B-36 bomber, which was a mammoth aircraft, powered by six Pratt and Whitney R-4360 piston engines, with the power absorbed by Curtiss Electric three-bladed pusher propellers, plus two General Electric J-47 jet booster engines under each wing.

Convair B-36 Bomber
Ref: http://en.wikipedia.org/wiki/Convair_B-36

The B-36 bomber design was initiated during WWII as a hedge against losing England to Germany. It was intended to enable the Army Air Force to fly bombing missions against Germany from the United States round trip non-stop. This aircraft flew 24-hour missions with an augmented crew (extra crewmembers). The crew referred to the B-36 as "six turning, four burning" for their engines (six propellers turning, and four jet engines burning). Apparently, most flights ended with at least one engine shutdown for one reason or another. Another interesting fact concerning the B-36 aircraft was it was crewed with an officer flight engineer. I thought I might be interested in becoming one, as I was going to be a degreed engineer. Alas, the B-36 was phased out before I was commissioned, and there were no more officer flight engineers.

AFROTC at Boston University
Instructors
<u>Captain Green</u>

Captain Green was a rated pilot who came from an assignment in an F-86D interceptor squadron in Japan. He bragged about being launched to intercept a Russian Bear bomber off the coast of Japan, with a wingman. His wingman aborted the mission with mechanical problems. Captain Green bragged about how his wingman was "chicken."

He was a stereotype fighter pilot; always pushing the limits. His wife had a baby, and he passed out cigars in class one day. Then he lit up and we all followed his lead. There was a No-Smoking policy in place at B.U. Someone in authority walked by and smelled the cigar smoke. Green was called on the carpet.

One day, he needed to use the Men's Room. There was a sign outside indicating it was closed for cleaning. He just used the Ladies Room. Green was in trouble again.

North American F-86D Interceptor
Ref: nationalmuseum.af.mil

Russian Bear long range bomber
Ref: en.wikipedia.org
Note the contra-rotating propellers that caused so much noise the aircraft could be heard miles away.

Major Knoxville

Major Knoxville was a WWII bombardier who had been recalled to active duty during the Korean War, and stayed in the Air Force to complete his 20 years for retirement. He seemed to be an intellectual, and probably would have made a good college professor. The major smoked a pipe constantly. After each bowl full, he ate a chocolate candy to coat his throat and lit up again. I hope he didn't end up with throat cancer.

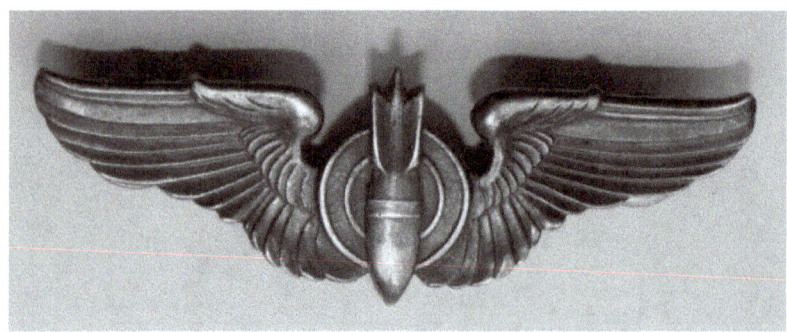

Major Knoxville wore these Bombardier Wings
Ref: en.wikipedia.org

AFROTC Advanced Corps

I previously discussed meeting the Advanced Corps Selection Board, consisting of the above active duty officers, among others, and made the cut for the advanced AFROTC program; I was matched against an Air Force future requirement for a navigator. Between my junior and senior year, I was required to attend an AFROTC summer camp, of four weeks duration. My orders were to attend the camp at Lockbourne AFB, Columbus, Ohio, in the summer 0f 1959.

Once in the advanced corps, I had the option of pledging the Arnold Air Society, a military (Air Force) fraternity. There was also the option of pledging Scabbard and Blade Society. The former was strictly for Air Force ROTC cadets, whereas the latter was for both Army and Air Force ROTC cadets. I elected Scabbard and Blade because the pledge duration was shorter, and I had little tolerance for hazing.

AFROTC Summer Camp

Between my junior and senior year in college, in August of 1959, I attended the mandatory four-week Air Force ROTC "Summer Camp" where the Air Force looked me over and vice versa. My camp was at Lockbourne AFB, near Columbus, Ohio.

Our group of B.U. cadets car-pooled to Lockbourne AFB in two cars. One cadet used his 1951 Chevrolet and another borrowed his father's 1953 Mercury. Upon arrival at the base, we were directed by signs to the designated parking lot. As soon as we parked the cars, there was someone screaming at us to double time to our barracks. This nonsense continued for the four week encampment. There were pushups for every minor infraction. My tactical officer was an active duty captain. He was a senior pilot, which meant he had 15 years of rated pilot service and a minimum of 2000 hours of flying time. Early in my training, he detected I wasn't putting out 100% during Physical Training (PT). He read me the riot act in order to get me to shape up.

Lockbourne was a SAC base, and had both relatively new B-47

bombers and KC-97F tankers assigned. As part of my orientation, I was required to fly on a KC-97 night Mass Gas (formation refueling) mission, which was interesting. We cadets attended the flight planning the day before the mission, and the briefing the day of the mission. The navigator on my aircraft sought me out, as a navigator future trainee, and said, "Why do you want to be a navigator? Don't you realize the Air Force just increased its commitment requirements from three years after school to four years?"

He was down on the whole concept, and obviously wanted to be released from active duty as soon as his commitment was up. I was just as resolute in wanting to become a navigator.

My Flight at AFROTC Summer Camp. My TAC officer is kneeling in the middle of the first row. The author is kneeling in the front row, the second from the left.

Skipping ahead, somewhat, I never saw my tactical officer again. Later in life, I heard through the grapevine he was a general officer, and he was on the first C-141 transport aircraft into Hanoi to pick-up the released POWs at the end of the Vietnam War. Recently, I came across his name again, and decided to research to see what happened to him. I found him in Wikipedia. He retired a four star general from the Air Force.

At summer camp, each cadet was required to participate in a 20 minute orientation ride in the back seat of a T-33 jet-powered trainer.

The Air Force photo of the author, on the next page, was sent to the local newspaper in Lynn, Massachusetts, my hometown.

Lockheed T-33 Jet Trainer
Ref: Wikipedia.org

Summer Camp Survival School

The Strategic Air Command (SAC) unit at Lockbourne AFB ran a survival school for the assigned B-47 bomber and KC-97 tanker aircrews. This school was located off-base in a rural area. Cadets were required to attend an abbreviated three-day course at this school. We were bused several hours into, perhaps, eastern Ohio or Kentucky and arrived at the encampment near sundown. As darkness approached, we were led into the woods. It was so dark I couldn't see my hand in front of my face. Each cadet put one hand on the shoulder of the one in front of him, in order not to wander off the trail. Upon arrival at the encampment, we were told by our instructors to post a guard and swap off every two hours. I was selected for the first guard duty, but didn't know whether to stand or sit; all I knew was I couldn't see and was certain nobody could see me.

The author posed as he was about to ride in a T-33 jet-powered trainer
Photo by Author

Suddenly, there appeared a bright light in the distance, arcing up towards me. It approached me and suddenly bounced off my chest. It was a flare. Yikes! Suddenly, the camp was surrounded by yelling and whooping instructors, as if they were our captors. We were warned to be more vigilant and the instructors faded into the woods again. How in the heck did they see me?

The next day, we had academics on how to survive in the woods. Following survival training, we were taught how to escape and evade (E/E Training). At the conclusion of this training, we were loaded into the rear of a dump truck and driven along a two lane dirt road into the woods. Periodically, several of us were dropped-off with a crude compass and hand-drawn contour map, and instructed to make our way to an objective. Unfortunately, my group consisted of city boys, myself included. We could not orient ourselves on the map, and certainly had no idea how to find the objective. The instructors gave us a two-hour head start, and then pursued us. If we were captured, we were to be taken to a simulated POW camp to undergo the "treatment." If we heard three shots of a shotgun, it was the signal the exercise was called off, and we were to make our way to the nearest road and wait for pickup.

We thrashed about, quite noisily, in the woods all night. In the morning we heard the shots in the distance. Eventually, we found a road and were picked up by the instructors. However, they cheated; the shots were fired if they couldn't find us, and we simply gave up. Off to the POW camp we went. Luckily, by the time we arrived, most of the exercise was over, and we had only a small taste of the B.S. Whereas, 10 pushups was the standard punishment for minor infractions at the base, 100 pushups, in the creek, were the norm at the POW camp.

At the conclusion of the survivor school, the instructors made a BBQ for the students. However, since we hadn't eaten for several days, our stomachs had shrunk, and we couldn't eat much. The steaks looked good, but we couldn't eat much.

Assessment of Summer Camp

I have been exposed to several AFROTC Summer Camps at bases where I was stationed. These encampments were a piece of cake compared to mine. For some reason, the Lockbourne AFB encampment was unduly harsh. Someone was screaming at us every waking moment. I felt the active duty cadre actually broke the morale of the cadets, and it was the wrong thing to do. It took some time to get over this experience.

As an aside, while at Lockbourne AFB, I was impressed with the B-47 radar maintenance shop. It had a tail gun from the B-47 set up with its radar control, and as we walked around the room, the guns followed us, guided by the radar. I thought it was cool.

Engineer, Set Refueling Power

I was excited about the Air Force, but was disillusioned with the bull crap associated with training to become an officer. I endured, but I did not make a good impression on my tactical officer, as I was depressed and completely unprepared mentally and physically for the heavy physical training requirements. I blame the Air Force active duty cadre at Boston University for the lack of mental and physical preparation. These officers should have prepared us for what lay ahead at summer camp, but to my recollection they did not. I had no idea what to expect, and was shocked upon arrival at Lockbourne AFB!

On the Sunday of the only free weekend we had at summer camp, one cadet and I decided to thumb a ride into Columbus, Ohio. We walked out the main gate, and were picked up by an enlisted man who had a gorgeous blue and white 1955 Ford Convertible, equipped with glass pack mufflers and standard shift. I still recall the driver winding up the engine through the gears of his car with the melodious muffler sound. I fell in love with the Fords of that era and hoped to own one someday. There is more to the story later on.

Boston University Graduation

In 1960, in my senior year at Boston University Engineering School, my peers interviewed with major companies for entry-level engineering jobs. I was envious, as they received offers of $500/month to start, from companies such as Pratt and Whitney Aircraft Engines. This was good money in 1960.

The starting pay for a new 2nd lieutenant including base pay, subsistence and housing allowance was $355/month. If I added $100 for flight pay, it came to $455/month, still less than my peer graduating engineers. I was not completely happy with this situation, but accepted it with the knowledge I would be doing much more interesting work. What bothered me about the pay differential was I had to fly 2.5 hours per month to earn the flight pay. Without the flight pay, I was substantially behind my graduating peers in engineering school in pay.

Engineer, Set Refueling Power

I graduated Boston University in May with a Bachelor of Science in Aeronautical Engineering, and a few days later, in June 1960, I received my commission as a 2nd lieutenant in the Air Force Reserve. I was slated to attend Undergraduate Navigator School, but had to wait for a class assignment. I could not find an engineering job for just a few months while I awaited orders for a call to active duty, so I continued to work part-time at the Stop & Shop Supermarket, for $2.00 per hour. This was not bad considering minimum wage was $1.00 per hour back then, but I was now an engineer, and felt embarrassed working for this "lowly" wage.

I was bitter about working part-time as a clerk with an engineering degree, but I accepted it knowing it would only be temporary. During this time I met Marilyn, dated her, and proposed marriage. We were married January 1961, just before my call to active duty. My orders finally came calling me to active duty in February 1961. I sold my 1950 Mercury for $150 just before I left for active duty.

CHAPTER 2; BACKGROUND, WORLD SITUATION, SAC BUILD-UP, TRIAD

In the late 1950s, the cold war stalemate was in full force. The United States was not sure how many long-range bombers and missiles were in the Soviet Union (USSR) arsenal, and the only way to find out was to overfly this secretive nation and photograph whatever could be seen. John F. Kennedy ran for president in 1960 on the concept of a "Missile Gap" existing between the USSR and the U.S. (Later. it was found out the "Missile Gap" was a myth, but it got JFK elected.) What <u>was</u> known was the USSR had the atomic bomb, but the US didn't know how many it had or if it had the means to deliver it, both in long range bombers and missiles. This was before the days of satellites monitoring what was going on. Periodically, a USSR spy smuggled out a poor quality photograph depicting the latest enemy fighter, and this photograph was published in the latest issue of Aviation Week.

President Eisenhower had secretly approved a plan to recruit Central Intelligence Agency (CIA) pilots, from the Air Force, to fly extremely high altitude spy missions in the classified Lockheed U-2 aircraft over the Soviet Union, and photograph their missiles and aircraft parked on their airfields. It was believed the U-2 could fly significantly higher than any intercepting Soviet fighter aircraft. This belief held until luck ran out and the U-2 piloted by Frances Gary Powers was shot down in May 1960. This was just before I graduated from Boston University and received my commission in the Air Force Reserve. I remember reading the cover story in the New York Times about a high altitude weather recon aircraft going off-course and being shot down in the Soviet Union. I bought the story, but this was before I realized my government frequently lied!

As a matter of interest, the U2 flew so high, breathing 100% oxygen was not sufficient to move the oxygen molecules across the membrane in the lungs. Use of a pressure-breathing configuration solved this. Instead of breathing-in the oxygen, the pilot relaxed his diaphragm, and the pressurized oxygen system filled his lungs. He then forcibly exhaled. This was the opposite of normal breathing, and it must have taken some getting used to.

Imagine doing this for hours on end. Also, a preflight hour of pre-breathing of 100% oxygen was required to purge the body of the nitrogen molecules. This was necessary to prevent an occurrence of the "bends" sometimes experienced by deep sea divers.

Dwight Eisenhower was in his last year of his term as president, when Powers was shot down, and relations with the Soviet Union went stone cold. Just after John F. Kennedy (JFK) was inaugurated as president in January of 1961, he inherited from President Eisenhower a mature plan to instigate a revolt in Cuba by supporting an invasion by Cuban freedom fighters at the Bay of Pigs. Newly elected President JFK withdrew his support for air cover for this invasion, at the very last minute, and on April 17, 1961 the freedom fighters were defeated on the landing beaches, and were killed or captured.

Shortly thereafter, JFK and Soviet Premier Nikita Khrushchev met at a summit in Vienna, during June 3 – 6, 1961. The meeting was presumably to discuss the future of the divided Berlin. Khrushchev sized up his adversary, JFK, as a weak president, especially after the abortive Bay of Pigs debacle, and determined the U.S. would not respond to his construction of a wall in Berlin to separate East and West Berlin. Construction actually began August 13, 1961. JFK was concerned and ordered a massive buildup of SAC. It was directed to increase the percentage of its bomber and tanker alert force from 30% to 50%, ready to go to war with minimum warning! All this was going on while I was in Undergraduate Navigator Training on active duty in the Air Force.

It was a dangerous time. The U.S. had a large arsenal of nuclear weapons deliverable by the B-47 Medium Bomber, the B-52 Heavy Bomber, the B-58 Supersonic Medium Bomber, as well as Polaris missiles aboard submarines. The U.S. strategy to keep the peace at that time was Mutually Assured Destruction (MAD) backed up by the

Engineer, Set Refueling Power

Triad of Weapon Systems. This Triad of Weapon Systems consisted of:

U.S. Navy nuclear submarines with Polaris solid fuel missiles under the sea

U.S. Air Force Minuteman and Titan missiles in silos across the Midwestern U.S.

U.S. Air Force manned bomber and tanker forces all over the world

B-47 medium bomber (2042 total produced by Boeing)

B-52 heavy bomber (744 total produced by Boeing)

B-58 medium bomber (116 total produced by Convair)

KC-97 tanker (816 total produced by Boeing)

KC-135 tanker (803 total produced by Boeing)

Fifty percent of the bomber and tanker alert force was on alert, 24 hours a day, so there would be no chance of destroying the entire retaliatory bomber and tanker force with a pre-emptive strike.

SAC Bombers and Tankers

KC-97G Tanker
Photo by Author

B-47 Bomber
Photo by Author

KC-135 Refueling B-52D, Internet photo
http://en.wikipedia.org/wiki/Boeing_KC
-135_Stratotanker

B-58 Hustler, Internet
Internet Photo
http://en.wikipedia.org

CHAPTER 3; KC-97 TANKER LINEAGE
Boeing B377

Much of the following description of the Boeing B377 came from the public domain, en.wikipedia.org.

The Boeing B377 Stratocruiser was a large long-range airliner, developed from the C-97 Stratofreighter, itself a derivative of the Boeing B-29D (renamed the B-50). The Stratocruiser's first flight was on July 8, 1947. The aircraft unique design features included two passenger decks and a pressurized cabin, a relatively new feature on transport aircraft. It could carry up to 100 passengers on the main deck plus 14 in the lower deck lounge.

The B377 was larger than the Douglas DC-6 and Lockheed Constellation and cost more to buy and operate. The aircraft reliability was poor, chiefly due to problems with four 28-cylinder Pratt & Whitney R-4360 Wasp Major radial engines and their four-bladed propellers.

There were five notable mishaps:

4/29/52 – Pan Am B377 crashed in jungle near Carolina, Brazil. Probable cause was separation of No. 2 engine/propeller, with consequent aircraft controllability issues. Fifty fatalities

3/26/55 – B377 Ditched Off Oregon Coast; 4 Fatalities

Propeller had separated causing severe control difficulties.

10/15/56 – B377 Ditched at Sea; San Francisco to Hawaii run

Aircraft lost power on Number one and four engines; there were no fatalities.

1/9/57 – B377 Disappeared at Sea; San Francisco to Hawaii run. There were 44 Fatalities. Unknown Cause

6/2/58 – B377 Manila; Aircraft had a hard landing which

resulted in a collapsed landing gear. Propeller contacted ground; a blade fragment killed one passenger.

A Primer on Hollow Steel Propellers

The author was a propeller installation engineer at Hamilton Standard from 11/65-11/69. The following is based on his personal experience: a propeller operates in a vibratory environment, due to the airstream inflow angle. Initially, the propellers on the B377 were the Hamilton Standard hollow steel-bladed design. These blades had a steel outer sheath forming the airfoil over "rubber" or foam to fill the void to the spar load-carrying member. Any nicks or gouges caused by impact with small stones, rocks or whatever, exceeding serviceable limits set by the manufacturer, during takeoff and landing, had to be blended out with a file to avoid stress risers. In a vibratory environment, the blade flexes as it goes through a revolution and the deficiency in the airfoil could eventually propagate into a crack. Left unblended, the crack could propagate until a section of blade separates from the propeller. Then the resultant unbalance easily could rip the engine right off its mounts. I am convinced this is what caused some of the accidents.

This variant of the famed B-29 bomber of World War II had improved Pratt and Whitney R-4360 engines, in place of the Wright R-3350 twin row 18-cylinder engines that were in the earlier B-29 models. The B377 had the same wing, tail, landing gear and engines, but incorporated a "double bubble" fuselage. Also in play was the fact the aircraft proved to be economically unfeasible for commercial service. This was due to the aircraft complexity. The Pratt and Whitney R-4360 aircraft engine was the last of the Pratt & Whitney Wasp family and represented the most advanced, powerful, but most complex piston engine powering a commercial aircraft. Its introduction came just before the first generation American designed jet-powered airliner became available, the Boeing 707.

Pan American Airways bought 20 of the total of 56 aircraft produced, for commercial service between San Francisco and Hawaii. The balance of the aircraft went in service with BOAC, United and

Northwest. The aircraft had a deplorable safety and reliability record. It suffered 13 hull-loss accidents between 1951 and 1970 with a total of 139 fatalities.

This propeller situation in addition to problems with the turbo-superchargers pretty much doomed the B377 to failure in passenger commercial service.

Pan Am Stratocruiser over San Francisco
Ref: en.wikipedia.org

B377 Boeing Factory Photo
Ref: commons.wikimedia.org

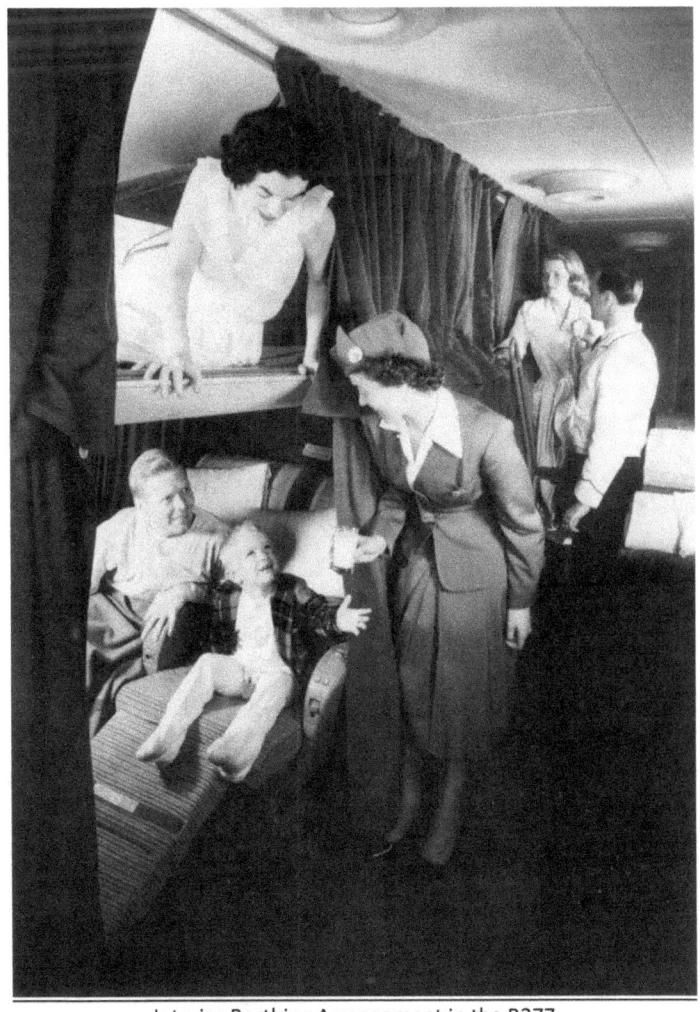

Interior Berthing Arrangement in the B377
By San Diego Air & Space Museum Archives -
https://www.flickr.com/photos/sdasmarchives/4590300860, Public Domain,
https://commons.wikimedia.org/w/index.php?curid=35287602

Boeing C-97 Stratocruiser
Boeing produced the cargo version of the aircraft, for the Air Force, and named it the C-97. It had special clam shell doors at the rear of the aircraft for loading cargo. The tanker version deleted these doors and substituted the flying boom pod for in-flight refueling.

Engineer, Set Refueling Power

Boeing C-97 Stratocruiser
Ref: commons.wiikipedia.org

SAC needed the KC-97 air refueling tanker (based on the B377 design) to extend the range of the jet-powered B-47 medium range bomber. With air refueling, the endurance of the B-47 aircraft was limited only by the physical endurance of the three-man crew (pilot, copilot and radar navigator). The KC-97 Tanker was pressurized, and heated (but not air-conditioned) and was powered by four Pratt & Whitney Wasp Major turbo-supercharged R4360-59B reciprocating radial engines.

The engine had a gear-driven supercharger in addition to a turbo-supercharger. This gear driven supercharger operated continuously, whereas the turbo-supercharger was brought into the picture by the flight engineer, by closing the engine exhaust waste gate, thereby routing the engine exhaust through a turbine. The turbine was coupled to a centrifugal compressor to more nearly simulate low altitude air density in the carburetor intake. Hamilton Standard four-bladed hydromatic, constant speed, variable pitch, full feathering and reversing solid aluminum bladed propellers absorbed the engine power. Early models had hollow steel propeller blades, which were subsequently replaced by solid aluminum blades, due to high maintenance requirements, as discussed above.

The KC-97 tanker was powered by piston engines, and as such was fueled by high octane aviation gasoline (115/145 Aviation Gas). The aircraft carried JP4 jet fuel in its fuselage-mounted air refueling tanks to refuel jet-powered bombers. The piston engines could not burn jet fuel, so the fuel tanks had to be segregated. In an emergency situation, the jet engines on the bombers could burn aviation gas, however. There was a "Manual Valve" on the main deck between the fuel tanks facilitating the transfer of aviation gas into the air refueling system. In order to preclude inadvertent transfer of JP4 jet fuel into the piston engine fuel system, the valve controlling this transfer could only be positioned manually.

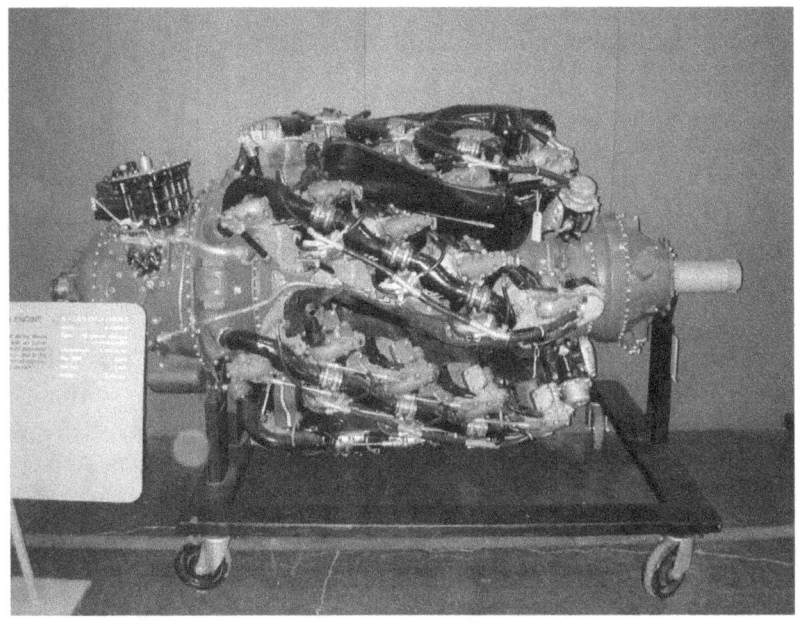

Pratt & Whitney R-4360 Engine
Internet Photo: wikipedia

Engineer, Set Refueling Power

Some characteristics of the R4360-59B engine were as follows:

Total piston displacement was 4360 cubic inches, ergo the R-4360 designation.

(Compare this displacement to one of the largest automobile piston engines, the 1976 Cadillac engine with its "massive" 500 cubic inch displacement.)

A total of 28 cylinders in a radial arrangement with four rows of seven cylinders, each row clocked to facilitate air-cooling of successive rows of cylinders.

Each cylinder had two spark plugs (224 spark plugs per aircraft).

ADI (Anti-detonate Injection Fluid) used at takeoff for improved power.

The engine was rated at 3500 Shaft Horsepower (SHP) @ 2700 Engine RPM.

The propeller was geared down by 0.375:1 to preclude the blade tips from reaching supersonic speeds. (Aerodynamic efficiency of the propeller blades falls off rapidly as supersonic tip speed is approached.)

The KC-97 aircraft was designed for 153,000 lbs. maximum takeoff gross weight (B377 Commercial Version was designed for 148,000 lbs. takeoff gross weight.) However, SAC needed maximum fuel offload to refuel B-47 aircraft and boosted maximum takeoff gross weight to the aircraft landing gear weight limit of 175,000 lbs. Needless to say, the KC-97 aircraft performance on takeoff at this weight was marginal. The KC-97 brake system made a peculiar noise when applied during taxi operations, somewhat resembling the "trumpeting" of an elephant, and when several aircraft were taxiing, it sounded like an elephant herd.

The Air Force used a crew of five to fly the KC-97; pilot, copilot, navigator, flight engineer, and boom operator. Also, as an aside, the engines were run in auto-rich mixture on the ground, which resulted in a heavy blue exhaust discharge. When a large quantity of KC-97s were taxiing, such as during an alert, the whole airfield turned blue with smoke!

CHAPTER 4; USAF UNDERGRADUATE NAVIGATION TRAINING (UNT)

Active Duty Call up Orders

After graduating from college and receiving my commission in the Air Force Reserve, in May 1960, I awaited a call-up to active duty with a class assignment to UNT, the 3567th Navigator Training Squadron. It took nine months for the assignment to come through. When the official orders came by mail, I knew instantly what was in the envelope. I was so excited.

Air Force Commissioning Photo; 2nd Lieutenant GR Dornfeld
Photo provided by Author

Transition from Civilian to Air Force

Marilyn and I were married January 22, 1961 in Roxbury, Massachusetts. My call-up orders were to report ten days later to James Connolly AFB in Waco, Texas to attend UNT. I bought airline tickets to travel ahead of Marilyn, on Sunday before my report date. On the previous day, Boston experienced a major snowstorm resulting in the closing of Logan Airport for much of Sunday. My flight was cancelled. I called a classmate from Boston University and a student officer in a class ahead of me, and asked who to call. He gave me the telephone number for the base commander. I called him at home, Saturday night. He instructed me to travel to Waco as fast as I could.

Then, I bought train tickets to travel to Chicago. Upon arrival, I booked a flight from Midway Airport to Dallas, Texas, and further to Waco. Upon arrival in Waco, I again called my classmate, as he volunteered to pick me up at the airport. He lived in the Bachelors Officers Quarters (BOQ) and was asleep when I called. My friend flew a training mission the previous night and slept late. In those days, the BOQ had one telephone on each floor, in the central hallway, and, when I called, nobody answered it. I took a cab from Waco Airport to the James Connolly AFB BOQ.

My classmate drove me around to look for a car. I found a 1955 Ford Crown Victoria in a used car lot and took it to a service station on the road to the base. It was owned by a retired sergeant who looked it over for me and told me not to buy it. I did anyway because I was young and stupid.

Ford to Chevrolet to Volkswagen
The Ford was indeed shot, and I traded it on a 1956 Chevrolet four-door sedan within a month. Two things I remember about the Chevrolet. First, the speedometer needle bounced as the car accelerated from a stop. The dealer told me, "They all do that." Right, Sure! Also, the Chevrolet used too much oil, although there was no visible oil leakage, and the exhaust was not blue, a sure sign of burning oil. Anyway, I had the garage that evaluated my 1955 Ford Crown Victoria rebuild the engine in the Chevrolet; it just needed piston rings.

A few months later I traded the Chevrolet on my first new car, a 1961 Volkswagen Beetle, in order to improve my gas mileage. I assumed after graduating from UNT, I was going to be assigned to Mather AFB for Radar Bombardier Training, in Sacramento, California and could use a car with better gas mileage. It was 1961 and JFK had boosted the SAC alert commitment from 30 to 50%, for reasons previously discussed. Every graduating class of navigators was assigned to SAC. I assumed I would be assigned to a B-47 bomber unit after graduating from the course at Mather AFB. One Saturday, we drove from Waco to a Volkswagen dealership in Dallas, and traded the Chevrolet on the spot. We drove the new Volkswagen home the same day.

Now I needed an apartment. My classmate heard of a modern apartment about to become vacant, as an upper classman was graduating. I looked at the apartment and took it immediately. It was located at 1107 Jefferson Ave., Waco, Texas. Next, Marilyn flew to Waco to join me. We settled in to our new apartment.

Military Compensation as 2nd Lieutenant
My pay as a 2nd lieutenant was $455.88/month, including $100/month flight pay. I had to fly a minimum of 2.5 hours per month to qualify for this flight pay. The government paycheck arrived once per month by mail. Marilyn waited for the mail to be delivered and rushed to the bank to deposit the check before the bank closed. In those days, there was no direct deposit, nor were there convenient banking hours. I believe the bank closed at 2:30 P.M. It was difficult budgeting during the month to make the money last.

Classmates
My classmates in Class 61-20 were all 2nd lieutenants, except for one, who was a 1st lieutenant. He previously had an assignment at Tachikawa, Japan. Since he was the senior student officer, he was designated as class commander. Another classmate had prior enlisted service time, and had recently completed officer's candidate school, so he was older than the rest of us. He had a two year old pink-colored Rambler station wagon with air conditioning, which was relatively rare in 1961. His car was the only one with air conditioning, as I remember.

The majority of the officers in my class were recent college graduates and newlyweds. For most of us, this was the first time we had a decent paying job. It was an exciting time. We didn't know if we were going to make the grade, and if we did, we didn't know what our operational assignment was to be.

After each phase of training was completed in the classroom, student officers applied their knowledge in the air using this aircraft. Note the multiple bubbles on top of the fuselage for celestial navigation training, and the large "chin" radome under the fuselage for radar navigation training.

My New 1961 Volkswagen

The 1961 Volkswagen wasn't equipped with a gas gauge as standard equipment. The car had a firewall-mounted lever, between the two front seats. When the car started "bucking," as it was running out of gas, the driver had to kick the lever over from vertical to horizontal. This fed approximately one gallon of gas from an auxiliary tank. But, the driver had to remember to reposition the lever vertical when refueling, or the next time the aux tank would be empty. Marilyn drove the car to pick me up at class when she was the first to experience the gas shortage situation. She came through, remembering about the lever. Another interesting fact concerning the 1961 Volkswagen was the air cooled engine was up-rated from 36 to 40 horsepower and the first gear of the transaxle had synchromesh. Finally the car could be shifted into first without grinding the gears, while the car was slowly moving.

066   070

HQ AIR RESERVE RECORDS CENT (ARC) United States Air Force 3800 York St, Denver 5, Colorado	EXTENDED ACTIVE DUTY ORDERS	DATE 26 Nov 60	SPECIAL ORDERS NUMBER A- 7427

BY DIRECTION OF THE PRESIDENT:

1. GRADE	2. RESERVIST'S LAST NAME – FIRST – MIDDLE INITIAL	3. AFSN	4. PAFSC
2nd Lt	DONEFELD, Gerald R.	AO 310 46 53	8611

5. COMPONENT	6.	7. YEAR OF BIRTH	8. SOURCE OF COMMISSION
Air Force Reserve	Initial /ADN/00 1	1939	AFROTC, Code J

9. DATE OF INITIAL APPOINTMENT	10. DATE OF SEPARATION	11. AERONAUTICAL RATING & FLYING STATUS
3 June 1960	Indefinite	Nonrated, not on Flying Status

12. HOME OF RECORD	13. TEMPORARY ADDRESS
8 Chase Street, Lynn, Massachusetts	

14. HAVING VOLUNTEERED FOR ACTIVE MILITARY SERVICE UNDER THE PROVISIONS OF Section 6d(1), Public Law 759, 80th Congress as amended by Public Law 51, 82nd Congress IS ORDERED TO EXTENDED ACTIVE DUTY Voluntary Deferment IN THE GRADE OF 2nd Lt ~~FOR A PERIOD OF~~ in Career Reserve Status

ASSIGNMENT

15. IS RELIEVED FROM RESERVE ASSIGNMENT

Hq ConAC (IRS) ARRC

ON DAY PRIOR TO EFFECTIVE DATE OF ACTIVE DUTY. EFFECTIVE DATE OF DUTY: 1 Feb 61

16. ASSIGNED TO 3567th Navigator Training Squadron (ATC), James Connally AFB, Texas to attend Class 60-20N, Navigator Primary-Basic Course #153138. Duration of course is in excess of 20 weeks.

AUTHORITY:

Message, ATC ATPOP-PA 72137, 22 March 1960.

REPORTING DATA

17. OFFICER WILL REPORT TO COMMANDER

3567th Navigator Training Squadron (ATC) James Connally AFB, Texas not earlier than 0800 hours and not later than 1200 hours, 7 Feb 61.

GENERAL INSTRUCTIONS

18. SECURITY CLEARANCE		19. ITEMS ON REVERSE SIDE WHICH APPLY
[X] NAC COMPLETED 31 Oct 58	FILED OSI-IG DD NR 4	7, 3, 6
[ ] BACKGROUND INVESTIGATION COMPL ___	FILED OSI-IG DD NR ___	

TRANSPORTATION

20. PERMANENT CHANGE OF STATION CHARGEABLE TO
5713500 148-103 P531.3 5503725 0200, 0300

21. TDY CHARGEABLE TO (Except for overseas & AD processing)

22. TRAVEL BY PRIVATELY OWNED VEHICLE WITH 6 DAYS TRAVEL TIME AUTH. IF PRIVATELY OWNED VEHICLE IS NOT USED, TRAVEL TIME WILL BE TIME OF COMMON CARRIER USED.	23. ITEMS ON REVERSE WHICH APPLY 13, 16, 17

ADDITIONAL DATA (Use reverse for continuation of above items as necessary)

FOR THE COMMANDER:

DISTRIBUTION

SA
PLUS

AFROTC Det 355, Boston University,
15, Mass

J. W. SCHILDKNECHT
CWO, W-2, USAF
Assistant Director
Directorate of Administrative Services

Amended 10 Feb 61

ARRC FORM 75 (TEST) AUG 60   NOTE: ITEMS NOT CONTAINING ENTRIES DO NOT APPLY

My Active Duty Call-up Orders

My official Air Force Photo while at UNT
I do not know why I looked so forlorn.

Plantation Apartments at 1107 Jefferson Ave., Waco, Texas

Convair T-29 Navigation Trainer
Commons.wikipedia.org

Marilyn posed in front of our 1956 Chevrolet. The car was in the parking lot of our apartment complex at 1107 Jefferson Ave., Waco, Texas.   Photo by Author

Author's New 1961 Volkswagen Sandwiched between a Sports Car and a Late 1950s Oldsmobile Coupe. Photo by Author

Author in the Apartment Complex Parking Lot
Photo by Author

Life as a Student Officer

As a student officer, I was treated markedly different than a cadet at AFROTC summer camp. I was treated with respect, as a 2nd lieutenant, and enjoyed all the benefits befitting an Air Force officer. I joined the officer's club, which was a requirement in those days, and took advantage of the nightly bachelor supper specials. Marilyn and I ate for a dollar each and then attended the movie on-base for a quarter afterward. Basically, for a dollar and twenty five cents each, we ate out and attended the movies; not a bad deal. The officer's club had a dress code, back then. After 6:30 P.M. a coat and tie was required in the dining room. Before 6:30 P.M., casual dress was acceptable. We made sure we ate and were out of the club before 6:30 P.M.

Author's wife, Marilyn, posed in front of new 1961 Volkswagen. Marilyn was pregnant with Michael, our first born. Photo by Author

When I entered the celestial navigation phase of training, I had homework using the MA-1 Hand Held Sextant. I had to take sun and star observations (shots) from the back of the apartment complex with the sextant suspended from the outdoor clothes line.

MA-1 Hand Held Sextant
Reference Air Force Manual 51-40, Air Navigation

Course of Study at Undergraduate Navigator School
Listed below are the phases of study leading to graduation as a rated Navigator:

Physiological Training including an Altitude Chamber "ride"

Radio Navigation including Morse code

Map Reading

Weather

Deduced Reckoning (Ded Reckoning)

Celestial Navigation

Loran

Consolan

Pressure Pattern Radar Navigation

Grid Navigation

Operational Techniques

Physiological Training, including Altitude Chamber.
This phase introduced the student to the symptoms of hypoxia, or lack of oxygen, at altitude, oxygen systems on aircraft, rapid decompression in a pressurized aircraft environment, and the effect of white light on night vision. Training included an altitude chamber "ride" where the effects of hypoxia were demonstrated followed by a rapid decompression. The student navigators were taught the preflight test of the oxygen system. The acronym PDMcCripe was used to insure a thorough preflight was performed.

P   Check the Oxygen Pressure Gauge for 400-450 psi.

D   Oxygen Diaphragm; select 100% oxygen and blow back into the mask for 5 seconds. A breathing restriction needs to be detected.

M   Oxygen Mask – check for cuts, tears, creases, etc.

C   Oxygen Connection. Check the gasket on the male side of the hose. Pull the quick disconnect to ensure operation.

C   Oxygen connection and regulator; check regulator hose from quick disconnect to regulator for deterioration and damage.

R   Regulator; Check for freedom of movement of auto mixture knob. Check red "Emergency" knob for safety wiring in the "Closed" position.

I   Oxygen Indicator; Check the oxygen flow indicator "blinks" when breathing oxygen with the regulator set to the 100% position.

P   Portable unit; Checkout the nearest portable bottle.

E   Emergency cylinder (bailout bottle); check for the pressure gauge to read 1800 psi, minimum.

Map Reading, including the various chart projections, such as Mercator, Lambert Conformal, and Polar.

The earth cannot be sliced up and projected flat without distortion. Think of the earth as an orange; it cannot be sliced up and laid out flat. This phase of training identified the level and magnitude of the distortion on Mercator, Lambert Conformal and Polar projections, as well as developing skill at map reading.

Weather; this course provided instruction on high and low pressure systems, the wind directions around the two systems, wind gradients, 500 and 700-millibar weather maps, etc.

"Ded" (Deduced) Reckoning, using the plotter, dividers, and the mechanical computer.

This phase introduced the student to use of the driftmeter, basic navigation instruments, such as the compass, altimeter, and airspeed indicator. Double drift and groundspeed-by-timing methods for calculating the in-flight wind were taught. Also, position projection on the basis of the best known wind, or preflight wind was taught, a process known as deduced reckoning (Ded Reckoning).

A technique was taught for navigating around bad weather systems, called air plot. This consisted of keeping track of aircraft heading, airspeed and time, plotting the air position, and then applying the best-known wind component to obtain a fix. This technique was used when many alter headings were required to stay out of heavy cumulus cloud buildups, with their associated turbulence and precipitation.

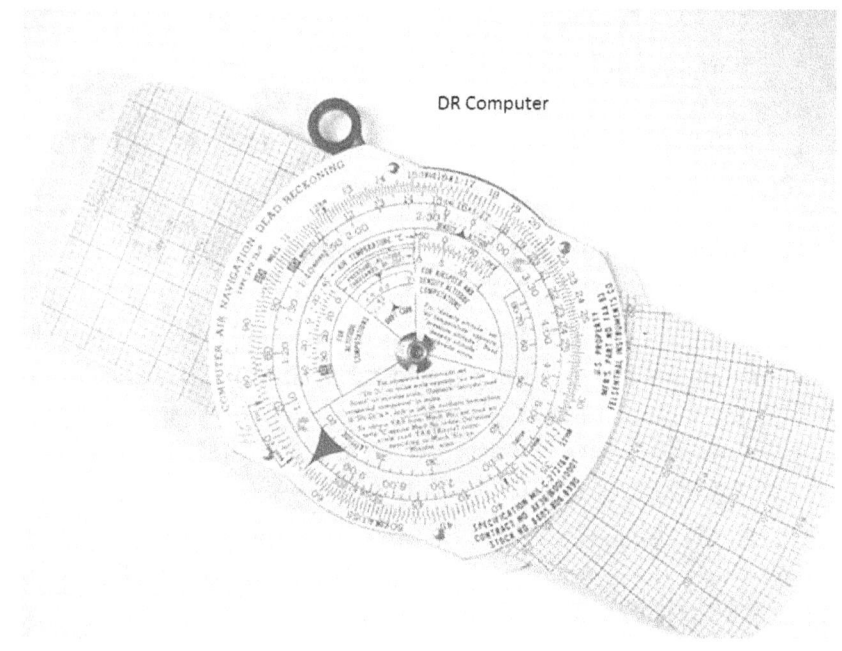

DR Computer

E6-B DR Computer
Wikipedia Photo

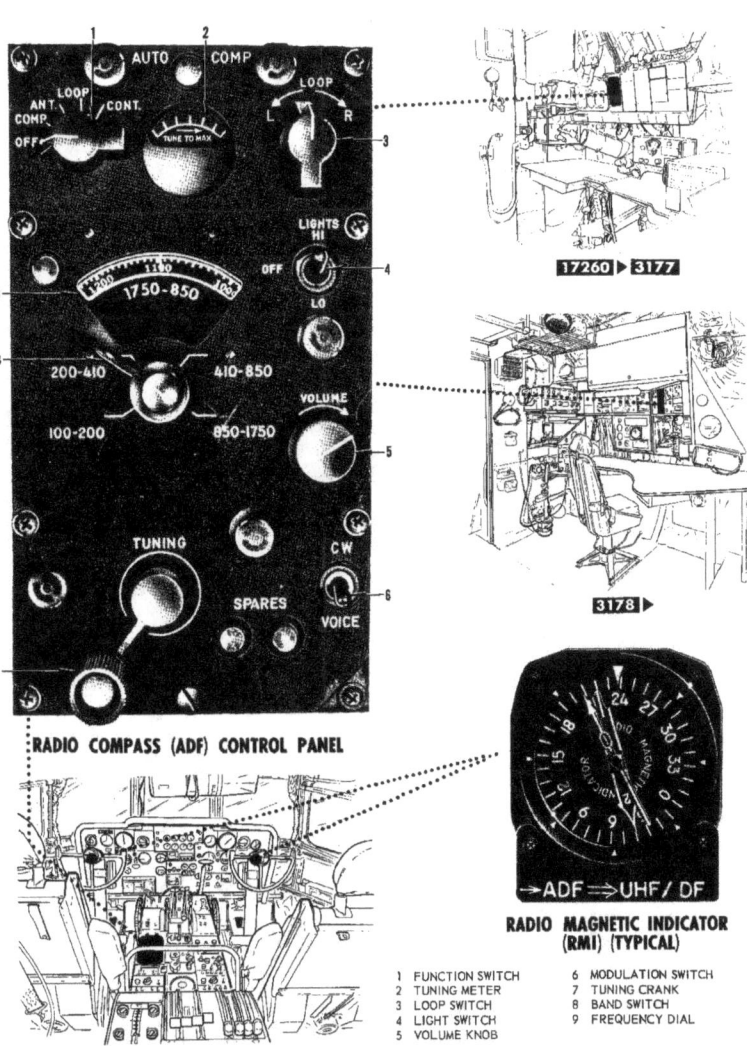

T. O. 1C-97(K)L-1

RADIO COMPASS (ADF) CONTROLS AND INDICATORS (TYPICAL)

Figure 4-21

1 FUNCTION SWITCH	6 MODULATION SWITCH
2 TUNING METER	7 TUNING CRANK
3 LOOP SWITCH	8 BAND SWITCH
4 LIGHT SWITCH	9 FREQUENCY DIAL
5 VOLUME KNOB	

ARN-6 Low Frequency Radio
Photo from KC-97 Flight Manual (T.O. 1C-97(K)L-1)

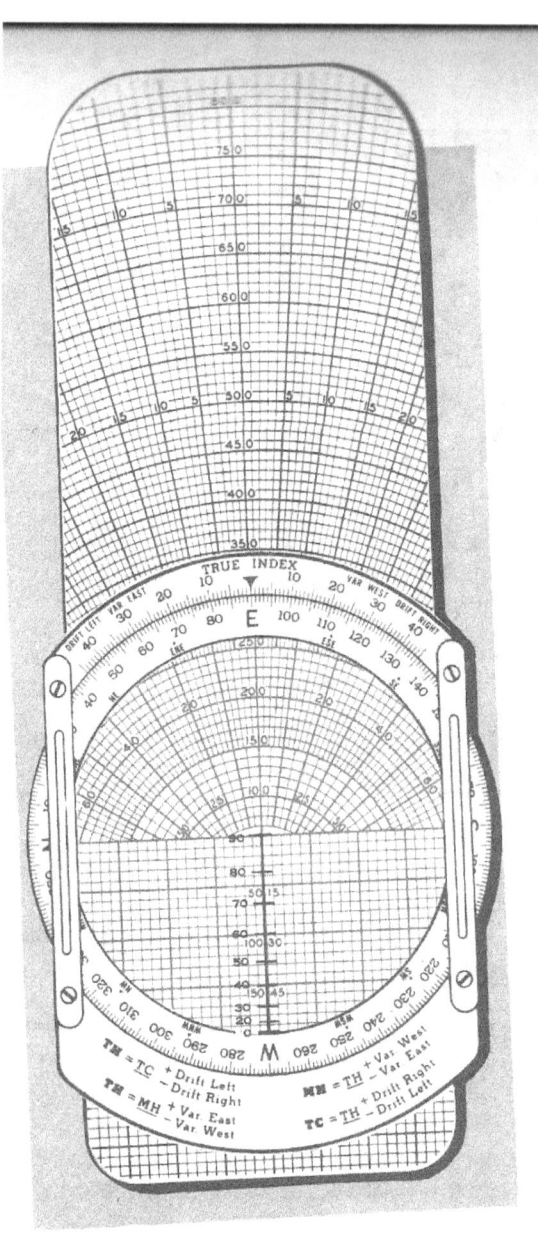

Wind Face Calculator
Reference Air Force Manual 51-40, Air Navigation

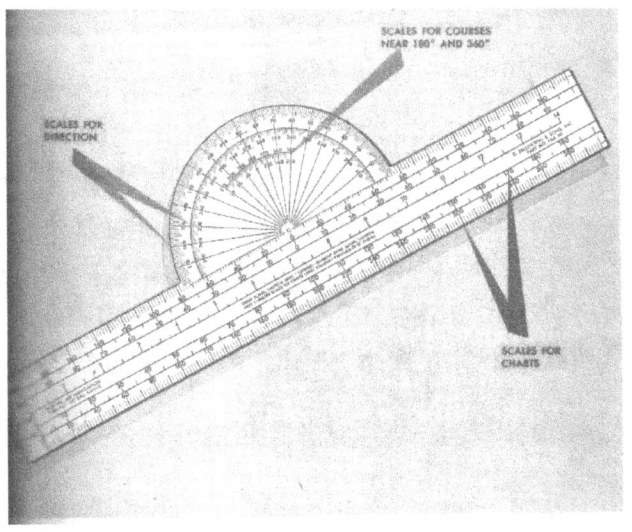

Wheems Plotter
Reference Air Force Manual 51-40, Air Navigation

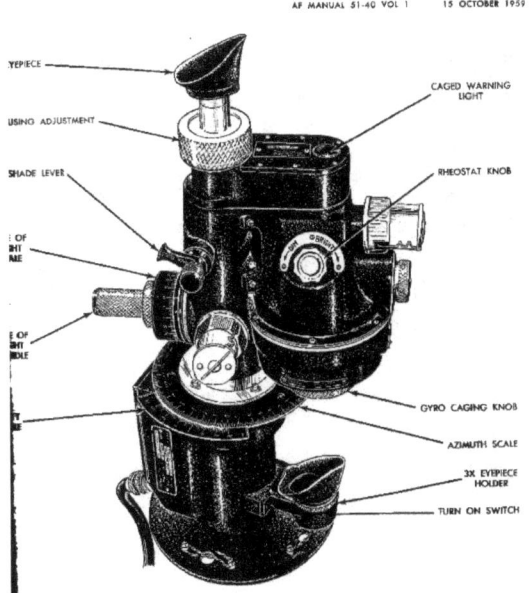

B6 Driftmeter
Reference Air Force Manual 51-40, Air Navigation

Engineer, Set Refueling Power

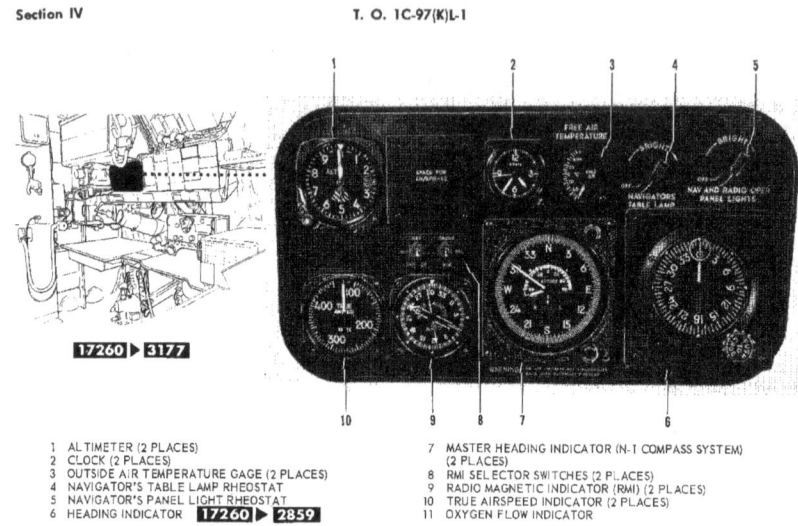

17260 ▶ 3177

1 ALTIMETER (2 PLACES)
2 CLOCK (2 PLACES)
3 OUTSIDE AIR TEMPERATURE GAGE (2 PLACES)
4 NAVIGATOR'S TABLE LAMP RHEOSTAT
5 NAVIGATOR'S PANEL LIGHT RHEOSTAT
6 HEADING INDICATOR 17260 ▶ 2859

7 MASTER HEADING INDICATOR (N-1 COMPASS SYSTEM)
   (2 PLACES)
8 RMI SELECTOR SWITCHES (2 PLACES)
9 RADIO MAGNETIC INDICATOR (RMI) (2 PLACES)
10 TRUE AIRSPEED INDICATOR (2 PLACES)
11 OXYGEN FLOW INDICATOR

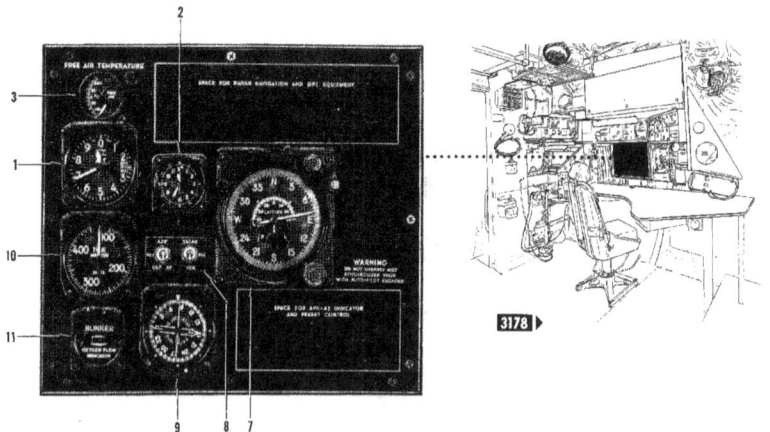

3178 ▶

NAVIGATOR'S INSTRUMENT PANEL

KC-97 Basic Ded Reckoning Instruments
Ref: KC-97 Flight Manual (T.O. 1C-97(K)L-1)

Engineer, Set Refueling Power

Celestial Navigation, including Precomputation, Sextant use, Sun lines and Night Three Star fixes.

Celestial Navigation is the process of assuming a position, and then calculating the altitude (height from the horizon in degrees and minutes) and Azimuth (angular orientation relative to true north) of the celestial body, using the HO-249 U.S. Naval Observatory Tables and Air Almanac. In the daytime, the sun and, if visible, the moon and/or venus could be used. At night, three stars giving cuts in azimuth of approximately 120 degrees apart in the sky would be used. Ideally, three bodies were used, but sometimes, only the sun would be visible.

Then, the altitude of these bodies would be physically measured using a bubble horizon in the sextant. The measured altitude, corrected for such factors as atmospheric refraction, motion of the aircraft and motion of the celestial body, etc., is compared to the calculated altitude, and a line is plotted from the assumed position, along and perpendicular to the azimuth. These lines or LOPs (Lines of Position) are adjusted for the time of the observation or "shot" along the aircraft projected course. Where the lines cross, the position of the aircraft lies.

An accurate watch was critical, and a time hack from the National Bureau of Standards, WWV, or the Canadian, CHU was obtained before the start of engines, usually at the crew briefing, but could be obtained using the aircraft radios.

All celestial observations were made using either the hand held sextant or the periscopic sextant, both of which had a two-minute mechanical averaging device. The "shot" was started one minute before the desired time and stopped one minute after. The navigator used the built-in mechanical averaging device to average the reading; this average sighting was displayed and used for the middle time.

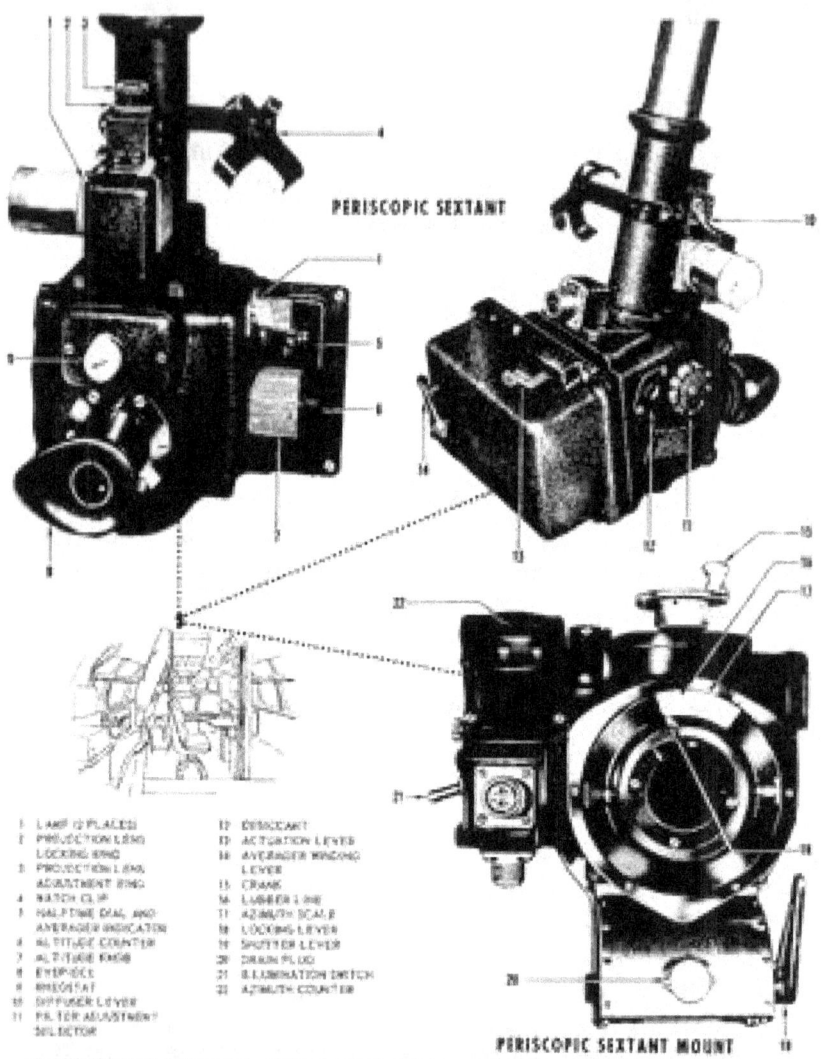

PERISCOPIC SEXTANT

PERISCOPIC SEXTANT MOUNT

1 LAMP (2 PLACES)
2 PROJECTION LENS
   LOCKING RING
3 PROJECTION LENS
   ADJUSTMENT RING
4 WATCH CLIP
5 HALF-TIME DIAL AND
   AVERAGER INDICATOR
6 ALTITUDE COUNTER
7 ALTITUDE KNOB
8 EYEPIECE
9 RHEOSTAT
10 DIFFUSER LEVER
11 FILTER ADJUSTMENT
   SELECTOR

12 DESICCANT
13 ACTUATION LEVER
14 AVERAGER WINDING
   LEVER
15 CRANK
16 LUBBER LINE
17 AZIMUTH SCALE
18 LOCKING LEVER
19 SHUTTER LEVER
20 DRAIN PLUG
21 ILLUMINATION SWITCH
22 AZIMUTH COUNTER

PERISCOPIC SEXTANT AND PERISCOPIC SEXTANT MOUNT CONTROLS

Figure 4-46

Ref: KC-97 Flight Manual (T.O. 1C-97(K)L-1)

Loran A, or Long Range Navigation, over water.

Loran A consisted of a system of Master and Slave Transceivers, located hundreds of miles apart, as a pair.

For example, Station 1H1 sent a Master signal triggering its Slave signal, when received. The difference in readings was preplotted on a chart as a series of parabolas. The Loran receiver on the aircraft allowed the two signals to be matched electronically by the navigator to determine the difference in time of receipt of both signals. The chart was preprinted with lines of constant time differentials. The navigator would then plot this difference in time, interpolating between the preplotted lines on the chart. Two or more station pairs allowed for a positive fix, where the plotted lines cross.

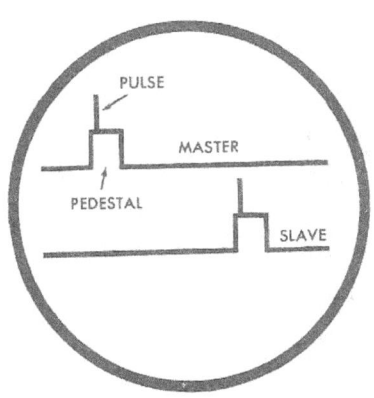

Pedestals under Pulses

Loran Signals Matched
Ref: Air Force Manual 51-40, Air Navigation

Engineer, Set Refueling Power

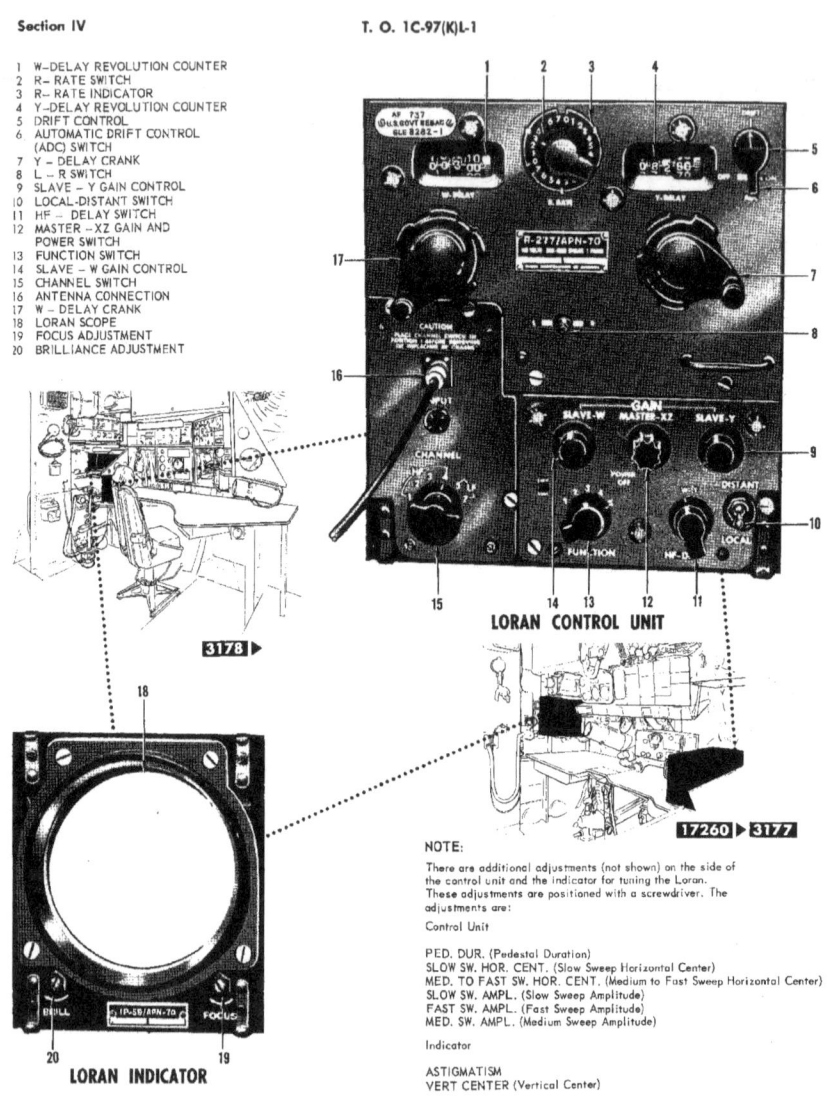

1 W–DELAY REVOLUTION COUNTER
2 R– RATE SWITCH
3 R– RATE INDICATOR
4 Y–DELAY REVOLUTION COUNTER
5 DRIFT CONTROL
6 AUTOMATIC DRIFT CONTROL
   (ADC) SWITCH
7 Y – DELAY CRANK
8 L – R SWITCH
9 SLAVE – Y GAIN CONTROL
10 LOCAL-DISTANT SWITCH
11 HF – DELAY SWITCH
12 MASTER –XZ GAIN AND
   POWER SWITCH
13 FUNCTION SWITCH
14 SLAVE – W GAIN CONTROL
15 CHANNEL SWITCH
16 ANTENNA CONNECTION
17 W – DELAY CRANK
18 LORAN SCOPE
19 FOCUS ADJUSTMENT
20 BRILLIANCE ADJUSTMENT

LORAN CONTROL UNIT

LORAN INDICATOR

NOTE:

There are additional adjustments (not shown) on the side of
the control unit and the indicator for tuning the Loran.
These adjustments are positioned with a screwdriver. The
adjustments are:

Control Unit

PED. DUR. (Pedestal Duration)
SLOW SW. HOR. CENT. (Slow Sweep Horizontal Center)
MED. TO FAST SW. HOR. CENT. (Medium to Fast Sweep Horizontal Center)
SLOW SW. AMPL. (Slow Sweep Amplitude)
FAST SW. AMPL. (Fast Sweep Amplitude)
MED. SW. AMPL. (Medium Sweep Amplitude)

Indicator

ASTIGMATISM
VERT CENTER (Vertical Center)

AN/APN-70 LORAN CONTROLS

Figure 4–26

APN70 Loran A Receiver
Ref: KC-97 Flight Manual (T.O. 1C-97(K)L-1)

APN-70 Control and Scope

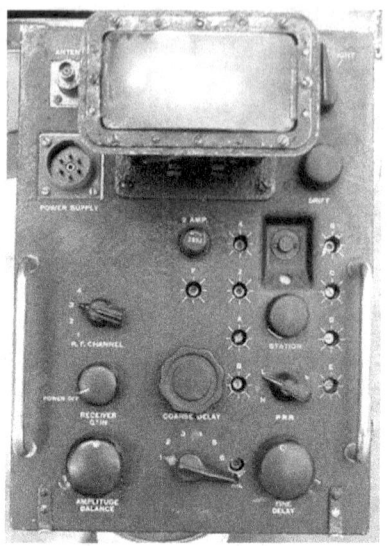

APN-9 Loran

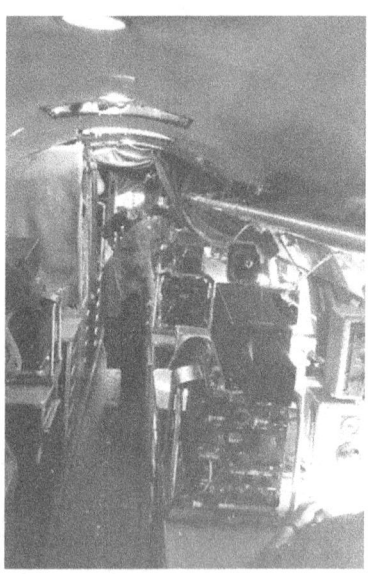

Student navigator stations on both sides of the T-29 aircraft.
In the foreground is an APN70 Loran receiver. Photo by Author

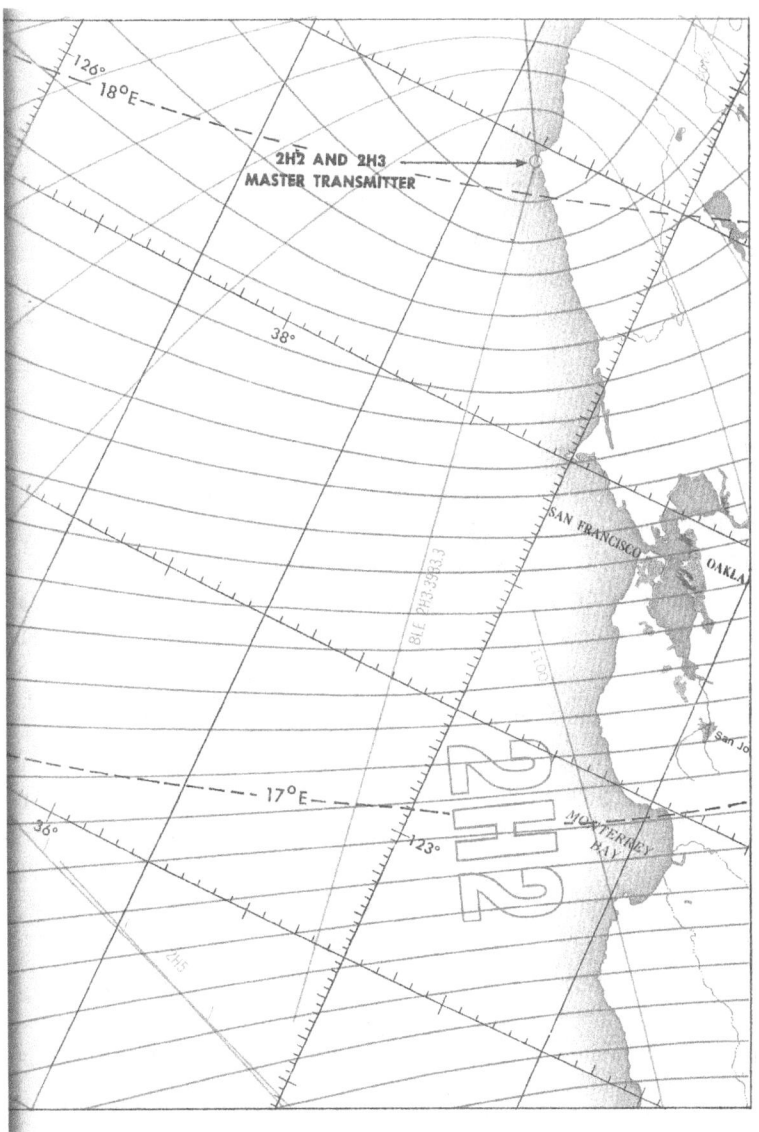

Loran A Chart
Ref: Air Force Manual 51-40, Air Navigation

Engineer, Set Refueling Power

Consolan

Consolan was a system developed by the Germans in World War II. Two phased signals were transmitted from a ground station at a coast. These signals were dots and dashes, and consisted of 60 characters. The signals were phased to each other, and, depending where the receiving aircraft was, the navigator heard the dots or dashes first. There would be a period where the dots and dashes merged into one, in the middle of the count. He counted the clearly audible first character and the second audible character. The total was subtracted from 60, and ½ the unknown were added to the first. This first total was plotted as a line on the preprinted chart.

Pressure Pattern

The aircraft flying into a higher-pressure area, in the northern hemisphere, drifted to the left of course, as the winds rotate clockwise around a high pressure area. The opposite was true in the southern hemisphere, and the directions were reversed flying into a low-pressure area. By keeping track of the difference in measured readings of the pressure altitude and radar altitude, and knowing the approximate groundspeed, the number of miles drifted left or right of the aircraft true heading could be determined and plotted on the chart.

This resulted in a "course" line, or a Line of Position (LOP) nearly parallel to the aircraft course. This line could be crossed with another LOP from any other navigation aid, such as a sunline or single Loran station, etc. The beauty of pressure pattern is it could be used to keep the aircraft on course, when the aircraft had no other navigational aids, such as over long expanses of ocean, when celestial navigation could not be used because of overcast conditions. The speed of the aircraft would not be known, but at least the aircraft would be on-course.

Radar Navigation

The airborne radar transmitted energy to the ground and displayed the reflected energy as targets on a scope with range marks and a bearing indicator. The range and bearing from the aircraft to the identified target is plotted as a reverse bearing, from the target to the aircraft. This navigation aid is invaluable over land, where reflected targets are easily identified, and when coasting-in from over water, but over long stretches of water, is of no use. The airborne radar is also useful in "Beacon" mode, where the system displayed the radar beacon reflected code; this was useful for flying formation (station keeping) and for rendezvous for air refueling.

Grid Navigation
Most aircraft were equipped with a gyro-stabilized compass system, called the N-1. This system was extremely accurate, to plus or minus ½ degree in heading. The sensor was located in the wingtip, and measured the horizontal component of the earth's magnetic field. However, at or above 20 degrees north or south latitude, the horizontal component of the earth's magnetic field became weak, as most of the lines of force were more nearly vertical, into the earth magnetic pole. Under these conditions, the compass system broke lock, and started to hunt or drift.

To facilitate navigation in these regions, an artificial grid was overlaid onto the chart that was not referenced to the magnetic field. The N-1 compass system could be put in "gyro" mode. The approximate latitude was set into the compass, and the aircraft was steered by the grid heading as measured on the chart, relative to grid north. The navigator's panel had a standby gyro and the pilot had one. Both of these gyroscopes were rated for precession over time, using celestial measurements. This was to allow the aircraft to be steered using grid, even when the weather obscured observation of the celestial bodies.

Operational Techniques
Techniques were taught, such as flying using the "Airway" system. In the United States and major countries all over the world, a system of navigation radio stations was located. A series of airways connect to these stations, similar to automobile roadmaps. This was how airliners navigated without human navigators.

Classroom followed by Flight Training
UNT consisted of classroom instruction followed by flight training in the Convair T-29 "Flying Classroom." A total of 472 of this aircraft

type were produced. It had twin Pratt & Whitney R-2800 engines rated at 2500 horsepower (HP), driving a three-bladed Hamilton Standard hydromatic, constant speed, full feathering and reversing propeller. The T-29 was equipped with stations for 14 student navigators and one radar station (for the instructor navigator). Each student navigator station consisted of a chart table, a cluster of basic navigation instruments (altimeter, airspeed indicator, radio magnetic indicator (RMI), and compass). Also, there was a driftmeter, APN70 loran set, SCR 718 radar altimeter, and sextant. In the roof of the fuselage were four astrodomes through which the student took celestial observations with the hand held sextant as well as one sextant mount for the periscopic sextant.

Each student navigator obtained a position every 15 minutes, in order to maximize the training for the flight time expended, using the navigation aid taught in that particular phase of schooling, during the four-hour training mission. His positions were scored compared to the position obtained by the instructor using radar. For example, during the celestial navigation phase of training, the student's three-star fixes were scored compared to the instructor's radar fixes. A total of approximately 100 hours of flying time were accumulated during these ten months of training.

We were not allowed to wear our flight suits off-base. Accordingly, there was a locker room where we dressed for a flight mission. I guess the administration didn't want the 2nd lieutenants walking around Waco in a cool-looking flight suit, trying to pick-up girls. After landing from a night training mission, I recall driving to my apartment, while listening to the radio. I was fascinated because the station was WMEX, in Boston. The signal was a skip wave off the ionosphere. It faded after a few minutes.

The class flew cross-country from Waco, Texas to Bermuda for a weekend, in order to practice both Loran and Pressure Pattern Navigation, both overwater aids to navigation.

T-29 Running Up Prior to Departing to Bermuda on a Cross Country Loran Training Mission

Graduating Class Assignments

Near the end of the course, a block of assignments came in from Air Force Personnel Center. The assignments were not by name, and each graduating navigator chose the assignment he wanted, based upon class standing. I was fourth in my class, so I had to wait until number one through three made their selections, while I hoped they would not select the assignment I wanted.

The assignments to Graduating Class 61-20 were as follows:

Quantity (1) to KC-135 Tankers at Turner AFB, Albany, Georgia

Quantity (1) to KC-97 Tankers at Dow AFB, Bangor, Maine

Quantity (1) to KC-135 Tankers at Plattsburgh AFB, New York

Quantity (1) to KC-97 Tankers at Otis AFB, Massachusetts

Quantity (1) to KC-135 Tankers at Griffis AFB, Rome New York

Quantity (1) to KC-97 Tankers at Little Rock AFB, Little Rock, Ark.

Quantity (1) to KC-97 Tankers at Selfridge AFB, Michigan

Quantity (7) to Radar Navigator School at Mather AFB, California

Quantity (6) to Electronic Countermeasures School at Mather AFB, California

Quantity (2) to Class Instructor at James Connolly AFB, Waco Texas

Quantity (2) to Radar Interceptor School at James Connolly AFB, Waco, Texas

The last assignment, to Radar Intercept Officer School, at James Connolly AFB, was not preferred, and graduates from the lower class standing were forced to accept it. The graduate became a radar intercept officer in the back seat of a hot jet aircraft. This was the good news. The bad news was the operational assignments for interceptors were mostly in remote parts of the country. The standing joke was you would love this assignment if you like to hunt and fish.

Photo of Graduating Class 61-20, Undergraduate Navigator Training
James Connolly Air Force Base, Waco, Texas; November 15, 1961
G Dornfeld is in the rear row, second from the right.
Air Force Photo Provided by Author

F-89 Scorpion used to train Radar Intercept Officers at James Connolly AFB, Texas
Ref: en.wikipedia.org/wiki/File:59fis-f-89-goosebay.jpg

After sweating out the selections by my classmates who were ahead of me by class standing, I selected KC-97 Tankers at Otis Air Force Base, Massachusetts. I lucked out and was able to get my preferred assignment. I was so excited I flunked my Radar Navigation check ride, which occurred on the same day. I did a makeup flight and passed with flying colors.

I was assigned temporary duty to Randolph Air Force Base, San Antonio, Texas, for three months, in order to attend SAC KC-97 Combat Crew Training prior to reporting to Otis AFB. Marilyn was pregnant with Michael, and I brought her to her parent's home in Revere, Massachusetts to have the baby, while I was at school. It was a stupid idea, as I could have brought her with me to Randolph, and just paid for our joint housing out of my own pocket.

There was just one instructor I remember from UNT. He taught us the Celestial Navigation phase of training, and he was a good instructor. He also was the only instructor who attended graduation ceremonies, and was genuinely happy to see us all pin on our wings. Just at this time, he received his operational assignment, as his one year tour as an instructor was up. He was so excited because he was going to SAC bombers, as he felt "real" navigators went to SAC. I wonder how long he felt this way with repeated week-long alert tours, but this is another story.

KC-97G Propeller Driven Air Refueling Tanker
Ref: commons.wikipedia.org

KC-135A Jet Powered Air Refueling Tanker Refueling a B-52
http://en.wikipedia.org/wiki/Boeing_KC-135_Stratotanker

At the completion of UNT, I received the shiny wings of a rated navigator and had accumulated 177 hours of flying time. I thought back to World War II, when graduating navigators had about the same amount of flying time, but were rushed into combat. I knew how green I felt, in peacetime, and could not imagine what it felt to be shipped directly to Europe or the Pacific to start flying combat, with this level of experience.

Navigator Wings

After graduating, I needed to fly 2.5 hours to receive flight pay for the month before I went on leave. I dropped-off my flight suit at the parachute shop. (The shop had heavy duty sewing machines.) This shop sewed my cloth navigator wings onto my flight suit, before the flight. I picked up my suit in a rush, without checking it. As I quickly dressed for the flight in the locker room, I didn't realize the shop had sewn the front of the suit to the back. As I inserted my leg into the suit, I ripped it. I felt embarrassed wearing the flight suit this way, but I had no choice.

CHAPTER 5; KC-97 TRANSITION

I took leave after graduation from UNT, and returned home to the Boston area for three weeks before reporting to the 4397th Air Refueling Squadron, Training, (SAC).

Upon reporting to Randolph AFB in mid-December 1961, I learned the class reporting date had been changed to early January 1962, so I had to hang around for a couple of weeks until school started. The school administrator told me he tried to contact me while on leave, to delay my reporting date until after the turn of the year, but he was unsuccessful in reaching me.

The school administration was worried the young student officers awaiting class initiation were going to use the down-time during the holidays to just take off and go wherever, so they ordered us to sign-in on the duty roster every morning at 0730. This meant I couldn't sleep late, even though I had absolutely nothing to do the whole day.

After a few days, I had enough of this nonsense, and got in the habit of signing-in late at night for the following morning. I never was caught. Nobody really gave a darn. The recalled pilots, who had an attitude because they were involuntarily recalled, refused to comply with the sign-in edict, and got away with it. Of course, the pilots were captains and I was a lowly 2nd lieutenant.

Recalled Pilots with an Attitude

The Berlin Wall had gone up the previous August, and world tensions were high. The Air Force had recalled pilots back to active duty, for assignment to KC-97 aircraft in SAC, and some of these pilots were classmates of mine. One in particular was upset about the recall, and walked around without his hat, which was called his "cover." His attitude was, "what are they going to do to him, recall him to active duty?" I doubt he signed-in every day the way the 2nd lieutenants did.

Ed Kostin had a Hardass for an Instructor Navigator

A good friend of mine and classmate from Navigator School, Ed

Kostin, was in my class. His instructor was a hardass. On Ed's first eight-hour KC-97 training mission, he was directed to perform a night celestial grid navigation leg. This was the most difficult type of navigation leg, and his instructor hit him up with it on his first flight. This navigation leg required the magnetic compass to be decoupled from the earth's magnetic field, and operate as a free gyro. Then a night celestial navigation leg was performed while the aircraft was steered using the gyro. After a grueling mission, the instructor navigator had Ed manually crank down the nose landing gear. Ed was wrung out after this mission.

He was subsequently assigned to the KC-97 squadron at Dow Air Force Base, Bangor, Maine. This was my backup assignment when we were selecting the assignments we wanted, by class standing, if my first choice was already selected. Ed was a regular officer, as opposed to my status as a reserve officer. In the early days of the Vietnam War, only regular officers were assigned to a one-year tour in Vietnam, in order to enable them to list combat on their personnel record, and thereby enhance their promotion potential. Ed received orders to Vietnam shortly after reporting to Dow AFB. He was killed crewing on a C-123 near the Cambodian border. Ed was the only member of my navigator class to be killed on active duty.

C-123K Aircraft
Ref: nationalmuseumaf.mil

Engineer, Set Refueling Power

We Flew Off The Chart!

I flew a training mission with my instructor navigator. We were number three of a six aircraft formation, called a cell. We performed the refueling of three B-47 bombers in a refueling track running north to south and ending at the Louisiana coast. The lead tanker instructor pilot wanted to give his student boom operator additional time practicing refueling a B-47, so he directed the formation to extend the refueling out into the Gulf of Mexico. The problem was, I had cut-down my chart to discard the excess I didn't need to fly this mission. The charts were of 1/4th of the U.S., each, so it was unwieldy without being cut-down. I explained the cut-down chart to my instructor, and was he pissed. He wrote me up for not covering all contingencies. I learned to carry an additional chart with me from then on.

Forging KC-97 Crew members out of New Navigators

Each flying training mission was planned for eight hours, and encompassed multiple navigation legs, emergency procedures, and refueling of B-47 bombers. The school trained me to be a crew member, not just a navigator. After an eight hour grueling mission, I was completely wrung out. There wasn't a minute to relax. The instructor navigator made sure I was performing at all times. The theory was, in SAC, if the navigator had time to eat his lunch, he wasn't doing his job. In contrast, in navigator school, we students were passengers until the aircraft leveled off and attained the departure fix we planned the mission around. Similarly, we closed out our log at the termination fix before descent and landing.

At KC-97 Combat Crew Training, we students navigated from aircraft brake release until landing rollout. As soon as the aircraft broke ground on takeoff, I tuned the radar set to paint ground returns from the small town of Fredericksburg, just off the active runway of Randolph AFB. On descent and into the traffic pattern, I followed the pilot using the approach procedures to back him up.

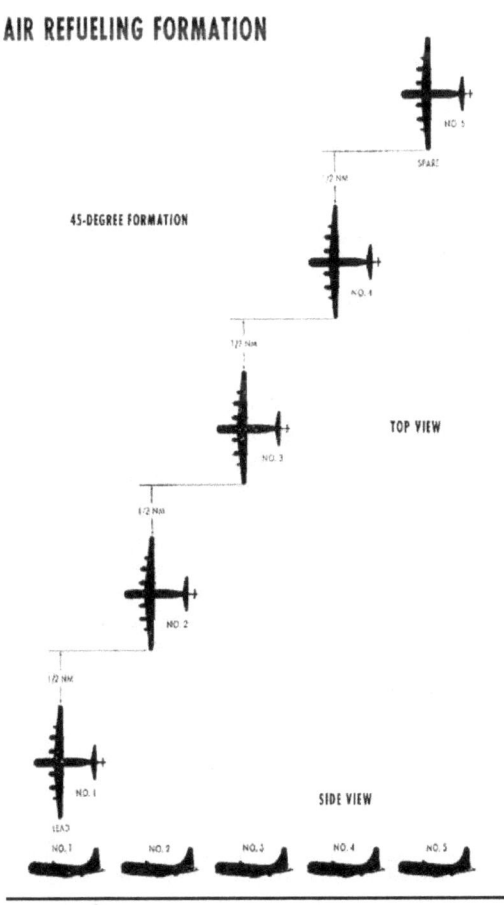

AIR REFUELING FORMATION

I learned the aircraft electrical system, hydraulic system, engines etc. I learned where every electrical power panel was located on the aircraft, and where on each power panel the circuit breakers were located for each of my electronic aids to navigation.

Additional Duties as a Crewmember

The navigator was integrated into the crew with additional crew duties besides navigation. He had the additional duty to crank down the nose landing gear, under emergency conditions, when the gear would not extend normally. (The boom operator had to manually crank down the main landing gear, under emergency conditions.) As an aside, whereas the navigator had to crank down, manually, the nose landing gear, against the wind, the boom operator had it somewhat easier, as the main landing gear cranked down with the wind. The navigator had to demonstrate the manual lowering of the nose landing gear once per year, during his annual check ride or "Stand Board" ride (standardization board consisted of the most experienced crew members who were responsible to check and maintain records of the proficiency of all the other crews.)

This required the navigator open the cockpit floor hatch, and lower himself into the lower forward electronics bay, affectionately known as the "Hell Hole" due to the higher temperatures in the bay. After checking in with the pilot over interphone the nose gear circuit breaker was pulled and the nose gear lever was positioned down, the navigator used a crank lever as a pry bar to move the gear clutch to free wheel position, and wait until the nose gear free fell part way, by gravity, but against the wind. When this was completed, the navigator cranked the gear the rest of the way down, approximately 220 turns. Then he re-engaged the clutch, the pilot reset the circuit breaker and the nose gear was recycled electrically to make sure it performed normally.

Another crew duty was to start the auxiliary power unit (APU) during aircraft descent. This unit was located in the lower forward compartment, near the lower forward entry door. The APU was a small jet engine built by Solar of San Diego, California driving a generator. The procedure was to position the lever closing off the intake and exhaust in pressurized flight, to the "Open" position, and engage and hold the spring loaded start switch until the unit started and attained 102% speed: He then repositioned the switch to "Reset," and then to "Run."

Engineer, Set Refueling Power

KC-97 APU Control
Located in Lower Forward Compartment

APU located just to the right of the APU Control
Photos by Author

When the APU was running, the little jet engine screamed. It was even loud when seated approximately eight feet away in the navigator's seat. I am sure this contributed significantly to my hearing loss in later life.

The navigator was designated as the primary crewmember to fight a fire in the lower forward compartment, specifically, a body heater fire, should one occur. There was a fire fighters mask with an oxygen bottle attached, on the cockpit bulkhead, along with a fire extinguisher bottle. Under a body heater fire condition, the navigator, when directed by the pilot, fought the fire. I have more to say on this in a later paragraph.

The navigator was required to verify the landing gear down locks and boom pin were pulled and stowed in the lower forward compartment, and was responsible for locking the lower forward entry door, all before the "Start Engines" Checklist. After the engines were started, and before the aircraft could taxi out of the parking space, the pilot rang the alarm bell to check its operability. The navigator responded over interphone, "Alarm bell loud and clear, down locks and boom pin stowed, lower forward entry door closed and locked, navigator ready to taxi."

The objective of SAC combat crew training in the KC-97 Tanker was to prequalify the crewmember for checkout in the aircraft, also known as the "Pre-Solo Check." I had my pre-solo check ride on March 9, 1962 in the KC-97F Tanker, tail number 251.

This temporary assignment forged members of my class into crew members in the KC-97 aircraft. First we went through ground school. Members of my class were force-fed all the technical details on the airplane. For example, I had to learn the location of all the electrical power panels on the aircraft and where on each panel the circuit breakers were located for each of my equipment, such as the APS 42 Airborne Radar, APN 12 Rendezvous Radar, APN9 Loran etc. In addition, I learned many technical details about the aircraft, such as symptoms of engine fires and type of fire, fighting fuselage fires, etc. As I was a degreed aeronautical engineer, I soaked up these details like a sponge, and was quite happy.

Life on Randolph Air Force Base
The Randolph AFB Officer's Club had a minority waiter who was rumored to have been there when current generals were 2nd lieutenants. He looked ancient. He would respond to our dinner order in the dining room by saying "Yes, Officer." It sounded more like, "Yes, Orrficerr."

I stayed at the Bachelor Officer Quarters (BOQ). There were two 2nd lieutenant roommates who were not navigators. These young newly-commissioned officers were wild. They had parties in the BOQ common area, and then retired to their individual rooms for an evening of sexual pleasure with their young female visitors, who probably were still in high school and lived on the base with their parents. In the room on the other side of the wall from mine, the lieutenant entertained his "date" by playing the guitar. When he moved out, it was discovered there was a "punch" hole in his wall. The Housing office wanted to blame me, but I explained it away.

It was winter in San Antonio. Some days were warm but there were also cold snaps. During my stay in the BOQ, the heating unit for the whole building became inoperative. The building comprised two floors and had, perhaps, 10 rooms on each floor. I figured someone would report the problem to the civil engineering office. The days dragged on, with no heat or hot water. I went to bed with every blanket I had, and I still was cold.

After about a week, I was so mad I made the call to civil engineering. I got a recorded message, and I blasted them for letting the situation go on for so long. Then I went to class. When I returned, the heat was operational. I took the longest hot shower of my memory, and boy did it feel good. Why nobody else called, I can't figure out. They sure outranked me.

I Studied with my Friend, Ed

The instructors compiled a loose leaf book with multiple choice questions about the aircraft. We were supposed to look up each question and answer it. Ed and I decided to divide up the questions and quickly complete the take-home test. An instructor caught us in the act and gave us heck. I suppose the instructor was right.

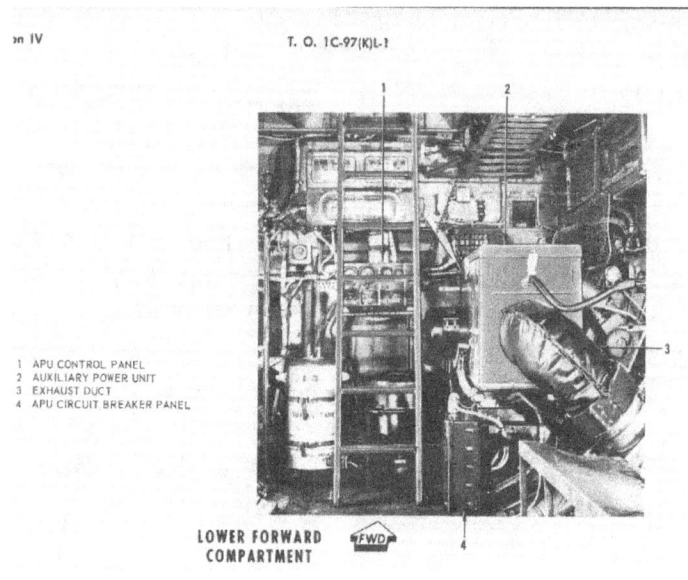

LOWER FORWARD COMPARTMENT

Auxiliary Power Unit
Ref: KC-97 Flight Manual

Flight Training Missions

Our flight training missions started in late January 1962. Students in the class were not just navigators. We had pilots transitioning into the aircraft, flight engineers and boom operators. Although each training mission was eight hours long, not all flights were to train me. I recall the instructor pilots demonstrating the flight characteristics of the aircraft to experienced student pilots. The instructor brought the engine power back to flight idle and raised the aircraft nose to the point of an incipient stall to demonstrate the stall characteristics. As the aircraft started to shudder, the instructor pointed the nose down to gain airspeed and preclude entering an actual stall.

My specific flight training was to perform two-hour navigation legs, usually celestial. Sometimes the navigation legs were back to back, so there was no relaxing in between. Then there was the rendezvous and refueling of B-47 bombers. We were taught the rendezvous procedure in UNT, but obviously never had to actually perform one. Now it was for real.

There were two methods for rendezvous, both using the Orbit. The primary method was for the navigator to turn on the radar beacon and put in the briefed coded settings. The bomber pilot relayed the ranges provided by his radar navigator and homed in on the tanker.

The other method was to use the APN12/76 rendezvous radar. My APN12 interrogated the APN76 on the bomber and painted a return on my scope. This way I controlled the rendezvous by giving the bomber headings and relaying ranges.

After arrival in the local area (of Randolph AFB) it was the pilot's turn to build proficiency with precision and non-precision approaches, missed approaches, touch and go landings, and landing with full propeller reverse thrust. During the landing pattern work, the navigator was required to follow the aircraft through the approaches using the approach plates.

After an eight hour flight, I was completely worn out. I did not have a minute to relax. The objective of the training at Randolph AFB was to prepare me to pass a pre-solo check ride, in preparation for assignment onto a combat-ready crew at my operational assignment.

KC-97G (SAC) Navigator Station
Photo by Author

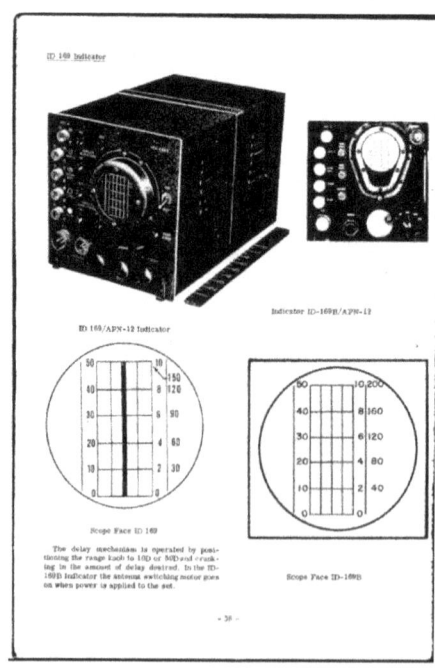

APN12 in the tanker aircraft

KC-97 Combat Crew Training
Randolph AFB, San Antonio, TX.
Jan 1962 – March 1962

Photos by Author

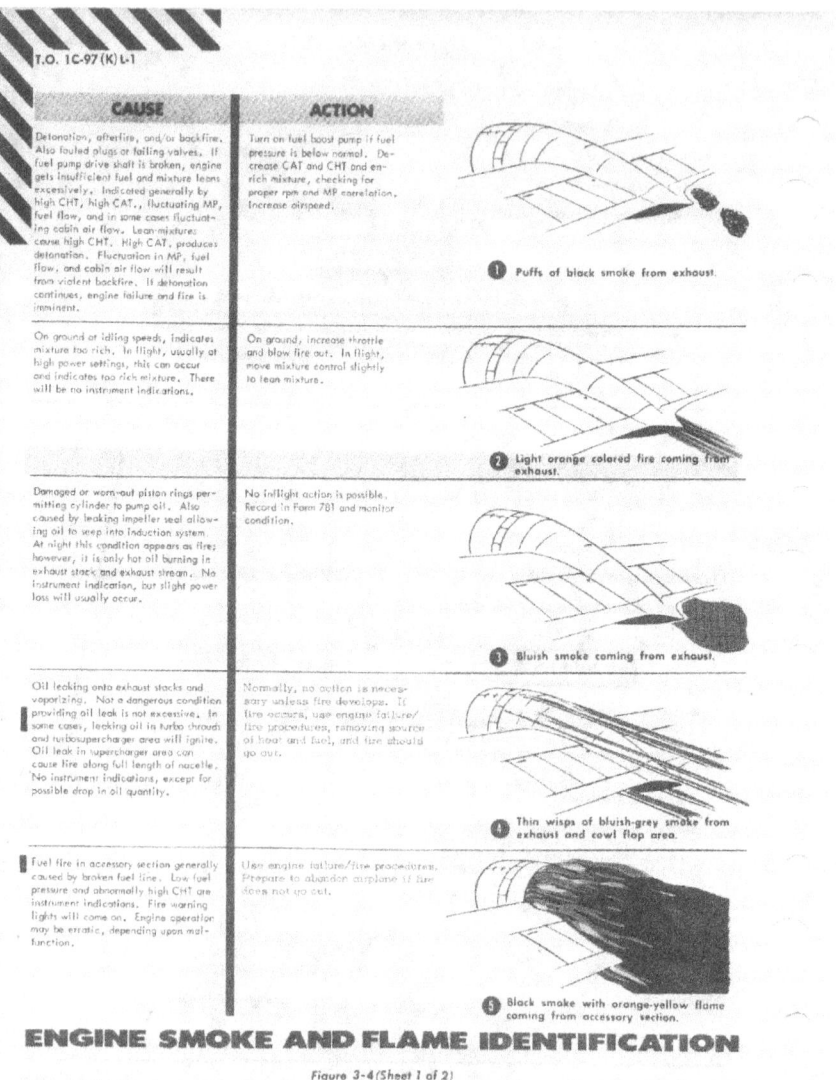

Engine Failure Recognition
Ref: KC-97 Flight Manual

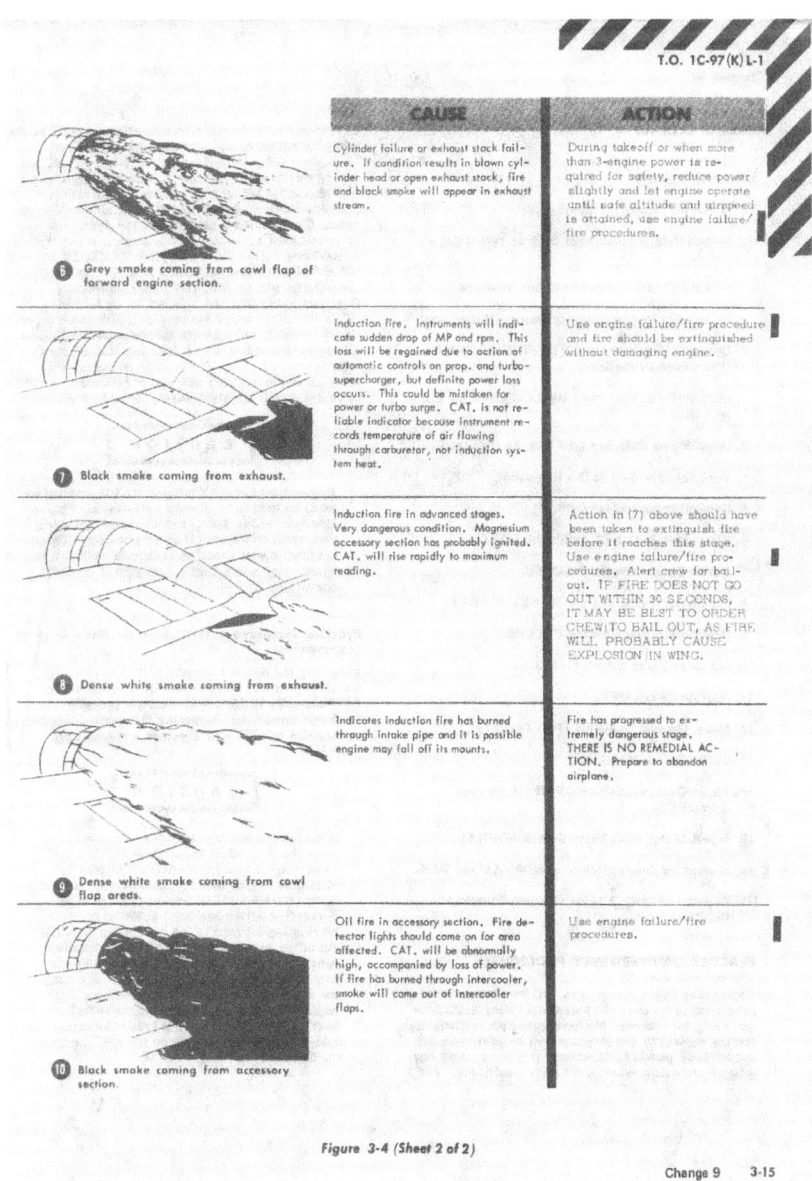

CAUSE	ACTION
Cylinder failure or exhaust stack failure. If condition results in blown cylinder head or open exhaust stack, fire and black smoke will appear in exhaust stream.	During takeoff or when more than 3-engine power is required for safety, reduce power slightly and let engine operate until safe altitude and airspeed is attained, use engine failure/fire procedures.
Induction fire. Instruments will indicate sudden drop of MP and rpm. This loss will be regained due to action of automatic controls on prop. and turbo-supercharger, but definite power loss occurs. This could be mistaken for power or turbo surge. CAT. is not reliable indicator because instrument records temperature of air flowing through carburetor, not induction system heat.	Use engine failure/fire procedure and fire should be extinguished without damaging engine.
Induction fire in advanced stages. Very dangerous condition. Magnesium accessory section has probably ignited. CAT. will rise rapidly to maximum reading.	Action in (7) above should have been taken to extinguish fire before it reaches this stage. Use engine failure/fire procedures. Alert crew for bailout. IF FIRE DOES NOT GO OUT WITHIN 30 SECONDS, IT MAY BE BEST TO ORDER CREW TO BAIL OUT, AS FIRE WILL PROBABLY CAUSE EXPLOSION IN WING.
Indicates induction fire has burned through intake pipe and it is possible engine may fall off its mounts.	Fire has progressed to extremely dangerous stage. THERE IS NO REMEDIAL ACTION. Prepare to abandon airplane.
Oil fire in accessory section. Fire detector lights should come on for area affected. CAT. will be abnormally high, accompanied by loss of power. If fire has burned through intercooler, smoke will come out of intercooler flaps.	Use engine failure/fire procedures.

6 Grey smoke coming from cowl flap of forward engine section.

7 Black smoke coming from exhaust.

8 Dense white smoke coming from exhaust.

9 Dense white smoke coming from cowl flap areas.

10 Black smoke coming from accessory section.

Figure 3-4 (Sheet 2 of 2)

Change 9   3-15

Engine Failure Recognition
Ref: KC-97 Flight Manual

KC-97L (Ohio Air National Guard) Navigator Station
Photo by Author

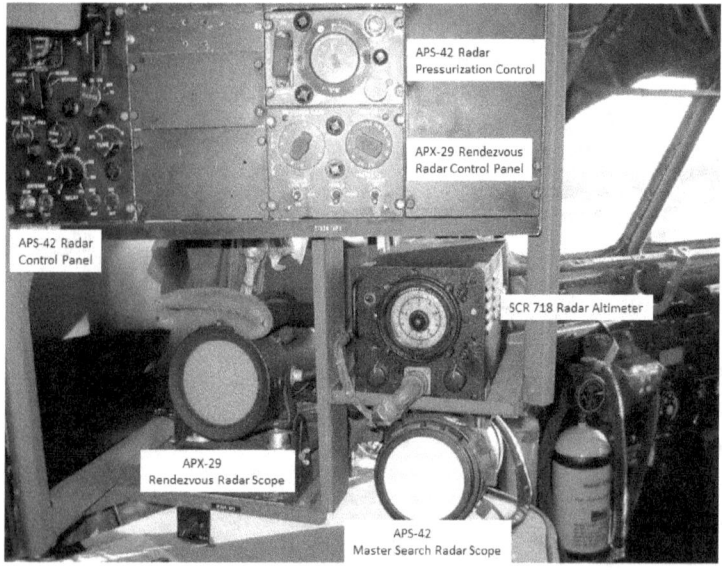

KC-97L (Ohio Air National Guard) Search Radar and Rendezvous Radar Controls and Scopes
Photo by Author

Engineer, Set Refueling Power

Compatibility Problem between the KC-97 Tanker and B-47 Bomber

The crew on the B-47 consisted of aircraft commander (AC or pilot), copilot and radar navigator. This medium range bomber required air-to-air refueling in order to travel intercontinental distances. With air refueling, the bomber range was limited by the physical endurance of the crew. In the early 1950s, the only aircraft capable of providing this air refueling support was the piston engine-powered KC-97 tanker.

There was a performance compatibility problem with this pair of aircraft. The all-jet powered bomber operated most efficiently at altitudes of approximately 30,000 feet. The tanker was capable of only 15,000 feet with heavy loads.

The compatibility problem was solved using tactical air refueling procedures. The bomber descended to 15,000 feet for the refueling and climbed back up to 30,000 feet after refueling. This was not the most efficient way of doing business, but it made use of the aircraft assets available at the time.

The formatting airspeed was a problem. Typically, it was 215 knots indicated airspeed (KIAS). At this speed, the B-47 was barely above stall speed, but the tanker was "balls to the wall." The tanker set Maximum, Except Takeoff (METO) Power to attain the formatting airspeed.

Upon taking on fuel, the bomber weight, and therefore stall speed increased. As the stall speed approached the formatting airspeed, the bomber AC called the tanker on UHF radio and requested "another five knots." If there was any more power to be had, the tanker flight engineer pushed the throttles forward. Usually, there was no more, as the engines were set for Maximum-Except-Take-Off (METO) power, which was 2550 engine rpm and maximum inlet manifold pressure from the turbo superchargers. The tanker pilot then pushed the nose of the tanker over to initiate a 500 foot-per-minute descent to keep the airspeed up. On heavyweight fuel transfers, the refueling could be completed as low as 12,000 feet altitude. Then the bomber initiated a climb back to 30,000 feet. The fuel transfer rate was 5900 pounds per minute, maximum.

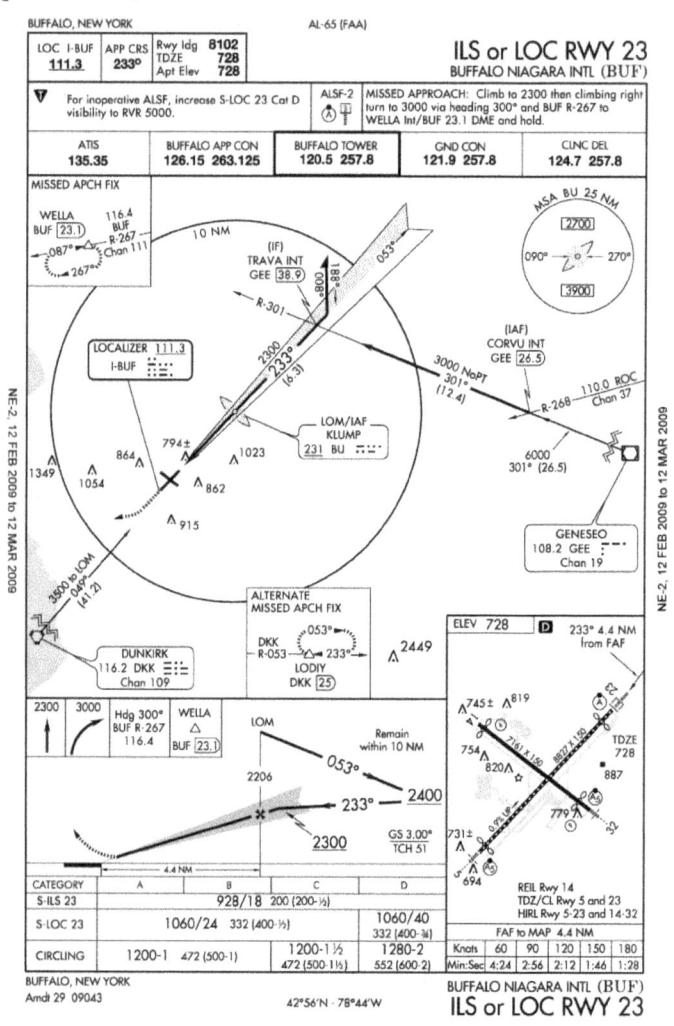

Approach Plate for Buffalo Niagara International

This is a typical Approach Plate

Engineer, Set Refueling Power

Boeing B-47 Bomber
Ref: commons.wikimedia.com

KC-97F Refueling a B-47 Bomber
This operation took place round the clock for years.
Ref: en.wikipedia.org

Engineer, Set Refueling Power

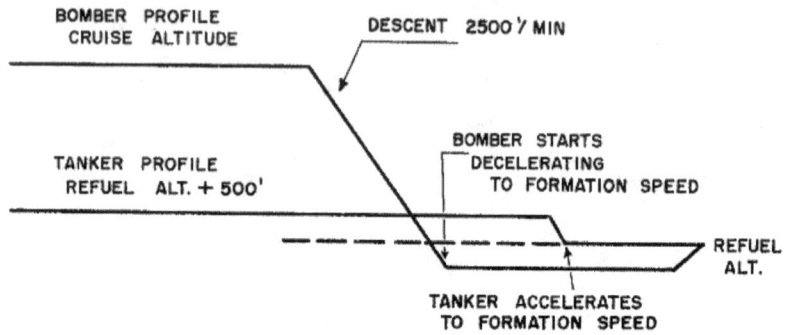

Side view of the rendezvous procedure.

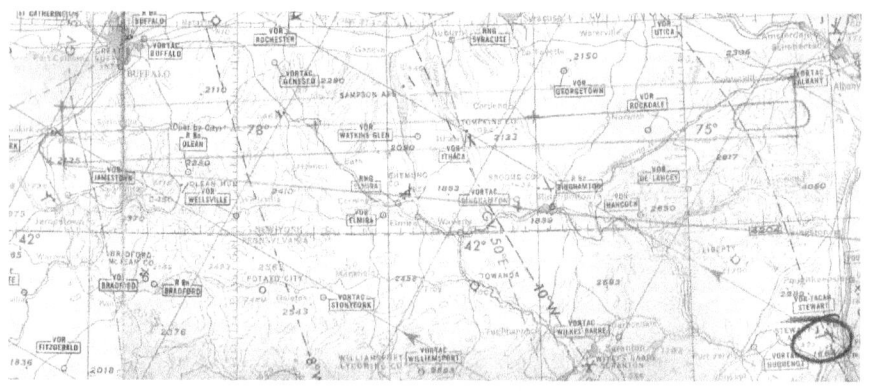

Tea Shop Air Refueling Track. This ran across the Finger Lakes of upstate New York.

Engineer, Set Refueling Power

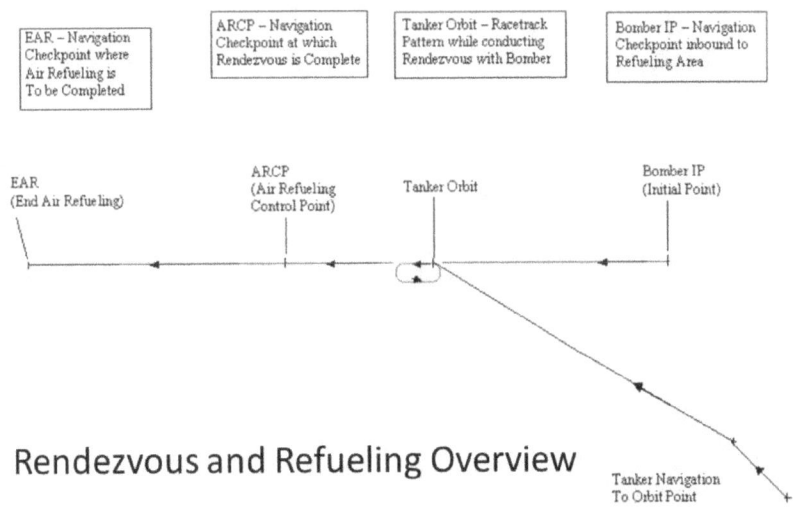

| EAR – Navigation Checkpoint where Air Refueling is To be Completed | ARCP – Navigation Checkpoint at which Rendezvous is Complete | Tanker Orbit – Racetrack Pattern while conducting Rendezvous with Bomber | Bomber IP – Navigation Checkpoint inbound to Refueling Area |

EAR
(End Air Refueling)

ARCP
(Air Refueling
Control Point)

Tanker Orbit

Bomber IP
(Initial Point)

Rendezvous and Refueling Overview

Tanker Navigation
To Orbit Point

Plan view of the rendezvous procedure

B-47 Bomber Taking on Fuel
Official Air Force Photo

Engineer, Set Refueling Power

Communication Security (Comsec)
Tactical call signs were assigned to each squadron and aircraft. This was to preclude the Soviet Union from tracking each aircraft, and its unit. Each squadron was assigned an identifier. For example, our squadron was temporarily identified as "Barn." Each aircraft in the squadron had a two digit number assigned. I might be flying in the aircraft assigned the number 21. So, our aircraft was Barn 21. The identifier was changed at irregular intervals, so next week the same aircraft might be "Weather 21."

Rendezvous Procedure
The tanker navigator directed the aircraft to the Orbit Point and established the Orbit pattern. This consisted of a four minute leg, a quarter standard rate turn to the left (four minute turn), another four minute leg followed by another four minute turn. The mission was planned for one complete circuit around the orbit pattern, or about 15 minutes. Once established in the orbit, the tanker AC attempted to make radio contact with the bomber on UHF radio refueling frequency. "Haystack 11, this is Barn 21, how do you read?"

Typically, the bomber was about 80 miles away, and its response was scratchy, as follows: "Barn 21, you are reading '3 by 2.'" Translation: Loud and Clear on a scale of 1 to 5. Three was medium volume and 2 was poor clarity. The tanker AC transmitted the weather in the refueling area and confirmed the fuel offload. Then the tanker AC turned the UHF radio over to the navigator for accomplishing the rendezvous. The navigator then became busier than a one armed paperhanger.

The tanker navigator turned the radar beacon to standby, and reported such to the bomber on UHF radio frequency. Then he turned the radar beacon to transmit. The bomber AC called, "Positive ID, 80 miles." The Bomber reported ranges every 20 miles down to 100 miles out, every 10 miles down to 10 and every mile thereafter. At the 30 mile range call, the tanker nav directed the pilot to depart the orbit for the air refueling control point. At the 12 mile range call, the pilot directed the flight engineer to set refueling power.

The Mass Gas mission used "Cell" procedures, which is another name for a formation. The mission consisted of two or more KC-97 Tankers flying to the refueling area, one mile in trail, or echeloned slightly to the right to avoid the wing downwash from the preceding aircraft. They were at level altitude, and the position was monitored by the navigator using the airborne radar (and the pilot/copilot on the repeater scope). Once the aircraft departed the orbit, the tankers echeloned out 45 degrees and stacked up at 500 feet altitude and one half mile separation. Typically, the bomber refueled from two tankers in order to get a larger offload.

KC-97 Lower Forward Entry Door   Air Refueling Tanks in Lower Forward Compartment, Looking Aft

Main Compartment Air Refueling Tanks, Aft Looking Forward
The Tanks are Stacked, One on Top of Each Other, on the Left Side
This leaves the floor available for cargo. Note the folding web seats
on the right side. The cockpit entry is forward.

KC-97 Crew Stations. Photos by Author

Pilot on left; Copilot on right

Navigator Station

Flight Engineer Station

P

Photos by Author

Air Refueling Tanks

The KC-97G had air refueling tanks stacked one over the other on the left hand side of the main deck of the fuselage, allowing personnel and cargo to be loaded on the main deck. There were tanks in the lower forward and lower aft compartments.

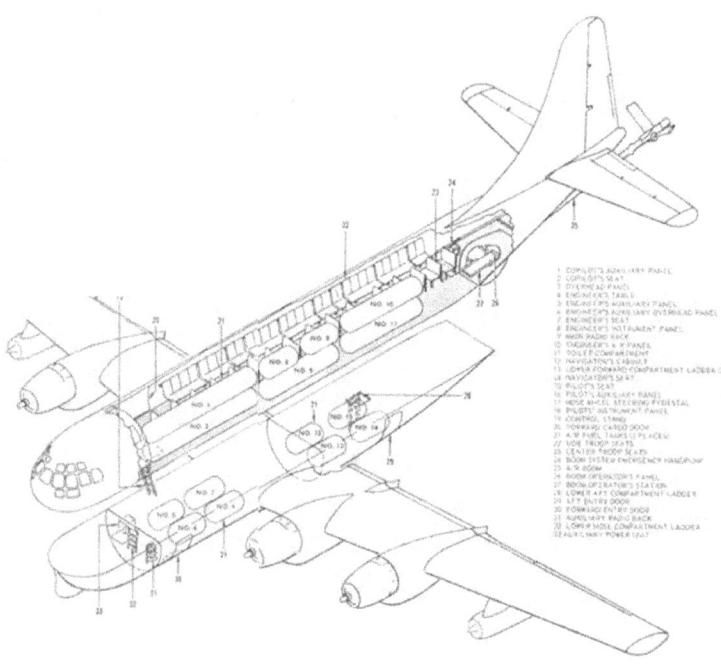

The location of the air refueling tanks on the KC-97G and L models
Ref: KC-97 Flight Manual

View of the main compartment, aft looking forward
Ref: amcmuseum.org

Air Refueling Tanks in the lower forward compartment
Photo by Author

CHAPTER 6; SAC OPERATIONAL EXPERIENCE

I reported to the 19th Air Refueling Squadron (ARS) at Otis AFB, Massachusetts in April of 1962. This squadron had been relocated from Homestead AFB, Florida, as part of the dispersal of the SAC retaliatory strike force in the late 1950s. The reason for the dispersal was to make it harder for the USSR to wipe out our retaliatory strike force with a preemptive strike. The 19th ARS was a tenant unit on Otis AFB, which was the home of the 551st Airborne Early Warning Wing, flying the EC-121 aircraft.

As an aside, I was jealous of the EC-121 unit navigators, as they were getting their maximum 90 hours of flying time per month, even though their flying was somewhat boring, flying search patterns in the Atlantic. SAC units (including mine) flew training missions to be prepared to go to nuclear war, in between alert tours, and crews were lucky to get 15 hours of flying time per month. This monthly flying time accumulation would require years before attaining the requisite 2000 hours to qualify for the senior navigator aeronautical rating, which I coveted.

I started flying training missions on May 11, 1962 leading to qualification as a crewmember. A humorous incident occurred during my first checkout in the KC-97, on May 14, 1962. I was yet to be assigned a crew, and was not scheduled to fly. The squadron operations office called me at home and informed me I was to fly the next day on a Mass Gas, with an instructor navigator (IN). I knew "Report Time" was one and a half hours before takeoff for the crew briefing. What I did not know was the "Report Time" for a Mass Gas was two hours before takeoff. I walked into the crew briefing room, on time, I thought. The formal briefing was already in process, and the room was completely full, with six crews in attendance. There were only two seats open, and both were up front! Embarrassed, I walked up front and sat down. Just then the squadron commander walked in, and saw I was sitting in his seat. I could have died. He was gracious, and slid over into the only other seat open, telling me to stay put. I looked for a hole to climb into.

Engineer, Set Refueling Power

My IN told me he was going to ask me one question; if I didn't get it correct, I would not be flying that night. The question was, "what was the correct GMT (Greenwich Mean Time)?" This was easy, as it was local time plus five hours. I passed.

On another mission I was scheduled to fly with an IN. The mission was to refuel a pair of B-47s from Pease AFB, New Hampshire. The takeoff was delayed due to dense fog at Pease AFB (There was dense fog at Otis AFB, also.) Ultimately, the bombers cancelled the mission due to weather. Our aircraft had the JP4 jet fuel downloaded, and more avgas loaded in order to fly a ten hour navigation training mission to get me checked out. I remember after a grueling ten hour navigation mission, upon descent, the pilot saying, "Hey nav, let's do a manual lowering of the nose gear." I was wiped out after that mission.

My Initial Check Ride

The time came for my check ride on June 25, 1962. Everything went according to plan during mission planning and preflight until we arrived at the aircraft after the mission briefing. The IN told me the instructor pilot (IP) wanted to accomplish a "Bravo" engine start. I had to admit nowhere in either KC-97 School or during any training mission had anyone mentioned what a "Bravo" start was. The IN was good about it, and said we would do it together, and didn't write me up for lack of knowledge. This IN doesn't remember this incident, but I have spent ten winters in Hawaii in retirement, and where this IN lives, and we have become good friends.

A "Bravo" engine start was an engine start from an alert configuration, where all the cockpit switches were pre-positioned for an immediate start. The navigator stood fireguard under the right wing until both number three and four engines were started (engines were numbered from one to four, starting from the left wing tip and going to the right, as viewed from the Pilot's seat.)

Once both engines were started, the navigator pulled the chocks on the right hand landing gear, and ran behind and underneath the aircraft in the rear to enter the lower aft entry door, climbed to the main floor and ran forward to his station in the cockpit.

EC-121 Airborne Early Warning Aircraft
Ref: Bwmoll3atEnglish wikipedia

KC-97 Aircraft from 19th Air Refueling Squadron just arrived at Sondestrom Air Base in Greenland for an alert tour. Photo by Author

Now back to the story. I stood fire guard for the engine start. Unfortunately, I didn't know where to stand, and I was exactly downstream of the exhaust outlet. The engine fired, and sprayed a fine mist of oil all over my flight suit.

I don't remember anything about the mission, so I must have done OK. I was checked out and assigned to a newly formed crew.

Promotion to 1st Lieutenant

I had been commissioned a 2nd lieutenant through the Air Force ROTC program in May 1960. My call-up to active duty was nine months later. I attended navigator school as a student officer and graduated in November 1961. Many of my peers on crews in the 19th Air Refueling Squadron were commissioned through the aviation cadet program, which meant they were a cadet while going through navigator school. My longevity for pay and promotion started when I was commissioned. Longevity for pay and promotion for my aviation cadet peers started when they graduated from "cadets."

Accordingly, my peers were already on crews as 2nd lieutenants when I reported to the 19th Air Refueling Squadron (19th ARS). I was almost immediately promoted to 1st lieutenant even though I was not yet checked out on a crew, whereas my peers from the "cadet" program were still 2nd lieutenants. I felt uneasy about outranking the other navigators, but this was the system.

Air Refueling Control Time (ARCT)

In SAC tankers, the critical time was the Air Refueling Control Time (ARCT). This was the scheduled time the tanker and bomber rendezvous was to be complete. The whole flight mission, both bomber and tanker, was predicated upon making good on this time. The flight mission was planned and takeoff time was backed out to arrive at the ARCT on-time. A five minute early takeoff time was planned, in case there was a maintenance issue with the airplane requiring repair. SAC scored each takeoff for an "On-Time Takeoff." Once, we had a failure of my airborne radar, while still on the ground. An electronics technician worked on the radar while the crew performed the engine checkout. As takeoff time arrived, the pilot decided to have the tech fly with us and fix the radar in route. Takeoff time was inviolate.

A B-47 bomber is in the "Observation Position." This was defined as rendezvous completion. This photo was taken from the KC-97 tanker boom pod, by the author.

View out the navigator's window. The number two engine is the closest, on the left.
Photo by Author

Backing-in a Night Celestial Navigation Leg

We had a newly assigned squadron commander of the 19th ARS. He flew as copilot with my crew on an orientation flight. Our mission was to refuel a B-47 bomber in the Fighting Fox Air Refueling Track which was oriented southwest to northeast over northern Maine. I never liked Fighting Fox because the winds over the mountains produced turbulence. We finished the refueling near the Canadian border and turned south to accomplish a night celestial navigation leg on the way back.

The procedure was to depart from a radio fix and steer the preflight computed heading using the best known wind. Since there was no opportunity to obtain an in-flight wind, I used the forecasted wind. Thirty minutes later, I took my first series of three-star "shots" with my sextant. As I plotted the results, the squadron commander called me on interphone and informed me we were getting too far off-course and I was to "get" on the radar to regain course.

Of course I had to comply, but now I had a problem; I was scheduled for a night celestial navigation leg and I was not able to accomplish it due to the direction of the commander. I spent considerable time backing-in the celestial data, after I landed, to "fake" a night celestial navigation leg. This was not the way it was supposed to be. What was I going to say to the scheduling officer the next day, if I did not complete my celestial navigation leg?

Once I was checked out as a crew navigator, I was on my own. I had a long way to go to feel comfortable in my role, but once there, it was terrific.

The author shooting the sun with the periscopic sextant
Photo by Author

SAC - The Primary Rendezvous Aid; the Tanker Radar Beacon

The primary means of rendezvous utilizes a radar beacon in the tanker, the APN69. This transceiver is triggered by the search radar of the bomber. The beacon return on the bomber radar scope represents the tanker position; range and bearing.

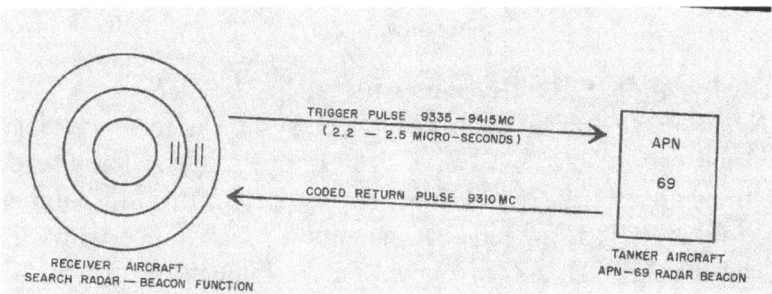

The receiver aircraft transmits on the unique radar beacon frequency. This triggers a response from the tanker's radar beacon.

RADAR BEACON CONTROL PANEL

Radar Beacon Control Panel in the Tanker. The code is set by positioning the pins in the lower row.

SAC - The Secondary Rendezvous Aid; APN12/APN76

The APN12/APN76 Interrogator-Responder system is the secondary electronic aid to rendezvous in SAC KC-97 operations. The tanker APN/12 transmits a trigger pulse which is received by the bomber APN/76. When it is triggered, it transmits a return pulse which is received by the APN/12 and displayed on the indicator scope in the tanker.

Engineer, Set Refueling Power

Indicator ID-169B/APN-12

ID 169/APN-12 Indicator

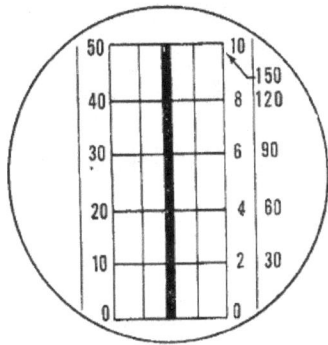

Scope Face ID 169

The delay mechanism is operated by positioning the range knob to 10D or 50D and cranking in the amount of delay desired. In the ID-169B indicator the antenna switching motor goes on when power is applied to the set.

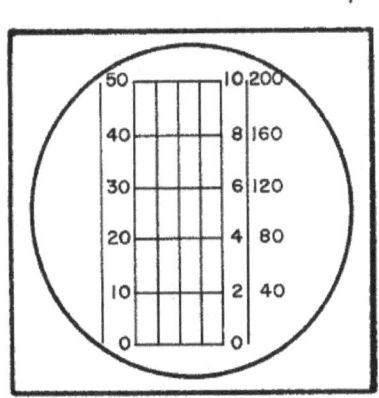

Scope Face ID-169B

Two different versions of the tanker APN/12 scope at the navigator station

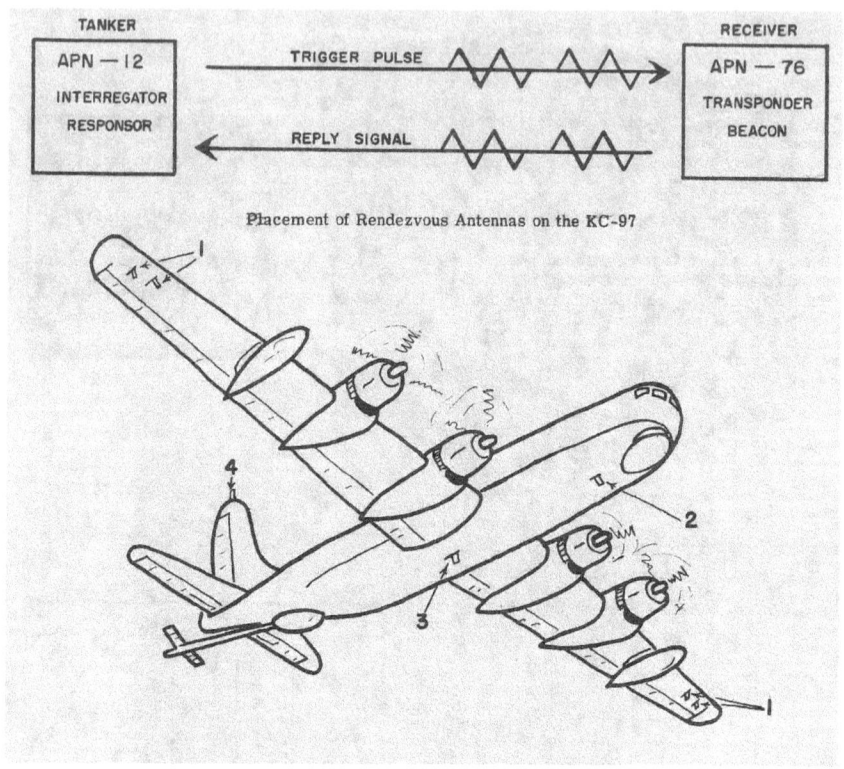

Placement of Rendezvous Antennas on the KC-97

My navigator checklist required I turn on my APN/12 set. But, rarely did I receive a signal from the bomber APN/76. Somewhere along the line, I came across the information the APN/76 set in the B-47 was behind the radar navigator. Apparently, he was concerned he could snag the ejection seat handle if he turned around to turn on the set. I don't know if this was true, but it seems plausible.

Only once did I accomplish a rendezvous using the APN/12-APN/76. Our aircraft was in the Orbit, communicating with a B-47. The bomber pilot told us his radar was inoperative, but he wanted to get-in the refueling. Suddenly, a signal appeared on my APN /12 set, and I was able to accomplish the rendezvous.

There were two more aids to a rendezvous. One was the UFH/DF. The bomber copilot transmitted on UHF radio. The tanker copilot positioned the selector switch to UHF/DF and the needle on the indicator pointed to the transmitting aircraft. No range was given, but a direction was indicated. Also, we had the option of firing a flare, which was visible by the other aircraft.

Integral Crew Concept in SAC
SAC used the integral crew concept. Our crew was formed, and identified as crew T-78, consisting of a pilot, copilot, navigator, flight engineer and boom operator. The pilot, copilot and navigator were officers, whereas the flight engineer and boom operator were enlisted. The crew did everything together; week-long alert tours (one week out of every three), deployments, flight training, ground training etc. Our mission in SAC was Mutually Assured Destruction (MAD). If the Soviet Union (USSR) was to launch a pre-emptive nuclear attack on the U.S., SAC bombers, their range extended by their tanker fleet, were to bomb the USSR back to the Stone Age.

SAC Alert Force
The 19th ARS had six tankers on alert 24 hours a day, 365 days a year. The aircraft were loaded to their maximum gross weight, as established by the landing gear weight limit. For the KC-97, this weight was 175,000 pounds. The crews rotated every seven days. The aircraft rotated on a different schedule. The alert facility, called the "Mole Hole," was located at the end of the active runway. Adjacent to the facility was the "Christmas tree," the hardened pads where the alert aircraft were parked.

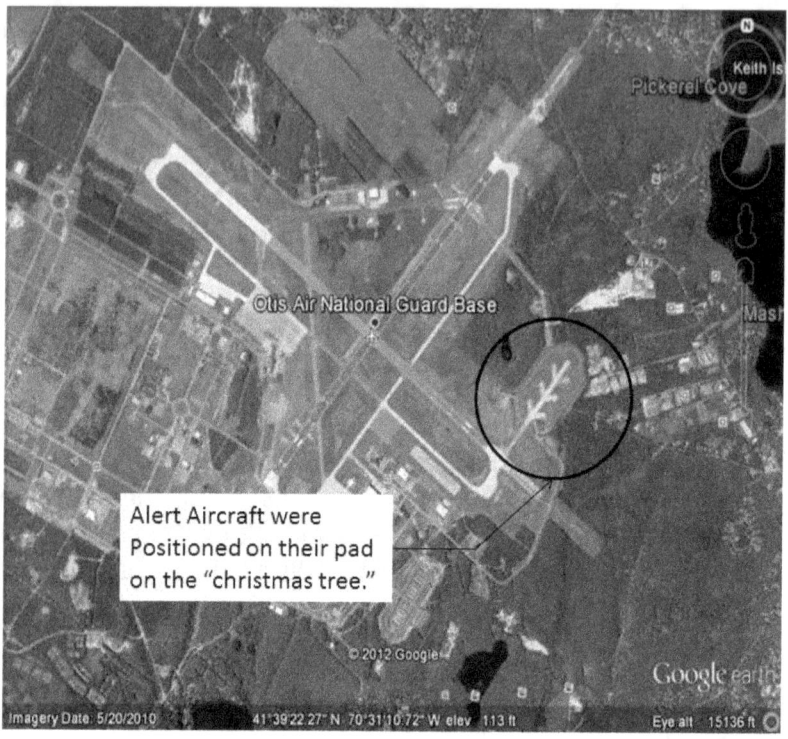

Christmas Tree at Otis AFB, Massachusetts. The Alert Facility was razed.
Ref: google earth

Alert

Crew change-over day was Thursday morning. Six fresh crews accomplished the Emergency War Order (EWO) study at 9:00 A.M. followed by a "tape" test. This was a test administered by a tape recorder to test the readiness to go to nuclear war. Failure to pass this test precluded the crew from assuming alert. A sample question on this test was as follows:

"If you are launched to go to war, and the bomber(s) required more fuel than was planned, what would you do?"

Correct answer, "Give the bomber all the fuel requested, even if it means the tanker cannot recover at the planned base. By "all the fuel" it is meant the tanker will give everything, allowing only sufficient fuel to clear the air refueling track. Then the crew was expected to bail out and to be on-their-own."

Following this test, each crew reported to its assigned aircraft to assure it was in the "Cocked" configuration, by running the "Cocked" Checklist. The crew about to be released from alert was already at the aircraft. This checklist called for all the switches and controls to be in position for an immediate engine start. The classified Emergency War Order (EWO) flight plan and charts were on-board the aircraft. The classified message verification material was on the navigators desk and adjacent to the copilot. In the event of an alert, once engines were started, the SAC Command Post transmitted a coded message requiring verification, using the classified material. Both the navigator and copilot had to verify the message individually, and agree it was a verified message, in order to launch the aircraft.

Practice Alerts

During the week-long alert tour, SAC Headquarters exercised the worldwide alert force by holding practice alerts. The klaxon blew in the alert facility, and all hell broke loose. The crews scrambled outside to their waiting vehicles. The first crewmember out the door from the alert facility was the vehicle driver, and he drove the crew the short distance to the aircraft. The pilot, copilot and flight engineer ran aboard the aircraft, while the navigator performed duty as the fire guard for the number three and four engine starts. Once these engines were running, he pulled the fire bottle to the right wing tip, ran back behind the running engines, fighting the propeller slipstream, and pulled the wheel chocks from the right hand landing gear. He then ran behind the aircraft and entered via the aft entry door. Meanwhile, the boom operator performed similar duties on the aircraft left side.

I was afraid I wouldn't be able to pull the chocks from the right hand landing gear in the winter, due to ice. Also, there was the possibility the landing gear tires could "pinch" the chock. Accordingly, before the number three engine started cranking, I pulled the chock. It didn't make a difference because the tires were so flattened due to the aircraft weight, there was no way the aircraft could move.

When all engines were running, and all crew members were aboard, the SAC Command Post at Otis AFB transmitted the classified message. This was decoded by both the navigator and copilot. The message directed the exercise as a practice Bravo or Cocoa alert.

Bravo meant just an engine start; the copilot called-in to the SAC Command Post the time of the completion of engine start. A Cocoa alert required the aircraft to taxi to the active runway and call-in the time as the aircraft crossed the runway threshold. Then the aircraft returned to the christmas tree and shutdown engines. Maintenance was called to fix any aircraft faults and the fuel tanks were topped off. During any week-long alert tour, there could be one or two practice alerts, or even none.

Minimum Interval Takeoff (MITO)

In the event of an EWO launch, the aircraft were to use MITO procedures. This utilized 15 second spacing between aircraft, in order to launch the alert force within the warning time of incoming missiles. This time was 15 minutes. The downwash off the wing of the preceding aircraft was a problem, and was particularly bad with heavier aircraft. The tactic used was for the lead aircraft to maintain runway heading after breaking ground; the second aircraft yawed upwind after breaking ground, the third yawed downwind, etc. The objective was to spread out the aircraft after breaking ground to stay out of the wing downwash of the preceding aircraft.

Every aircrew had to practice a MITO once per year, during a training mission. It was a nerve wracking event. This practice MITO was performed at moderate aircraft weights. On my last year on active duty, in 1965, SAC directed the annual practice MITO be performed at EWO weight. I was nervous. My shoulder straps were so tight I couldn't move. Thank G-d we came through it alright.

The Alert Facility (Mole Hole)

The Mole Hole was equipped with as much comfort as possible.

Engineer, Set Refueling Power

There was a lounge with a black and white TV, library, game room, cafeteria and crew briefing room on the main floor. Below grade were the sleeping accommodations, showers, etc.

The pilot had his own room, the navigator and copilot shared a room as well as the flight engineer and boom operator. Daily, after breakfast in the cafeteria, the crews assembled in the briefing room for a closed circuit televideo conference with the base weather station. The weather at Otis AFB as well as in the assigned refueling track was briefed, in case we went to war that day. Then the crews went to their aircraft to perform the "Cocked" checklist to insure the aircraft was still ready for immediate launch.

I was leery about having the klaxon blow while I was on the "john" or in the shower. Luckily it never happened with me, but it did with others. It was inevitable.

We were on an alert tour at Sondestrom AB, Greenland. The flight engineer, Sergeant Plowalski, requested the crew vehicle to go to the NCO club for lunch. He dropped us off at the base cafeteria where we ate. The rule was, in the event of an alert, the person going the furthest away from the flight line took the crew vehicle. He was supposed to pick up the rest of the crew and drive to the flight line.

Sure enough, the klaxon went off. It was an alert. The rest of our crew, the two pilots, navigator and boom operator walked out of the front door waiting for the flight engineer to pick us up. Meanwhile we observed other crew vehicles rushing past us. Soon, we heard the engines from the other aircraft starting.

Where the heck was Sgt. Plowalski? We were going to be in trouble if we busted the timing for engine start. Finally, he drove up. The pilot chewed him out for being tardy.

Sgt. Plowalski said, "Sir, I was taking a dump. I had to squeeze and wipe."

We made our engine start timing for the "Bravo" alert, but just barely. What happened to Sgt. Plowalski was what each of us feared would happen. Luckily, this never happened to me.

Home Alert vs. Reflex (Alert at a forward facility)

When I reported to the 19th ARS, the unit was tasked with alert at its home base, Otis AFB. Shortly afterward, the alert was shifted to Sondestrom Air Base, Greenland for a year. This was referred to as Reflex. The alert force was positioned further north, along the intended flight path of the bomber fleet. This allowed for attacking bombers to be refueled further north, while flying a great circle route over the North Pole to attack the USSR.

The alert facility at Sondestrom did not emulate the hardened mole hole at U.S. SAC bases. Also, the facility was further away from the flight line. Individual aircrew members were allowed to take the crew vehicle to, say, the NCO Club, as long as the vehicle was upstream of the route to pick up the rest of the crew when the klaxon blew, and the alert time could still be met. I "pulled" alert at Sondestrom AB every third week for a year.

Proficiency Flying

Our flight missions were designed to maintain proficiency in order to be ready to go to war. We flew day and night refueling missions, navigation training with an emphasis on celestial navigation, and the pilots accomplished their precision and non-precision landing approaches and landings. Unfortunately, we accumulated only 15 hours of flying time per month, on average. We "pulled" alert every three weeks.

Tanker Top-off Mission in England

The Pease AFB B-47 fleet had an alert commitment at Fairfield RAF Base, England. The aircraft were configured with external electronic countermeasure jamming pods (ECM Pods) on the outside of the fuselage. Since these pods presented more surface area to the relative wind, they resulted in more drag and therefore more fuel consumption. Normally, when the aircraft rotated home to Pease AFB, Portsmouth, New Hampshire, the ECM Pods were removed.

Engineer, Set Refueling Power

Someone figured out the manpower-intensive procedure of removing the ECM Pods could be eliminated by virtue of basing a KC-97 tanker at Fairfield RAF base. This tanker would take off early and top off the fuel tanks of the B-47 bombers after takeoff.

Critical Wind Factor
SAC required a critical wind factor be calculated before a flight plan was filed. This represents the greatest headwind component allowing the aircraft to make it to destination with the fuel onboard. In order to increase the critical wind factor for the rotating B-47 aircraft, the tanker topped off the fuel tanks after takeoff.

19th ARS assigned tanker top off mission
My crew was assigned this mission. Three B-47 bombers rotated home twice a week. On our first mission, we took off 45 minutes early and climbed out. The B-47 bombers started engines. We were in the soup at the assigned altitude of 15,000 feet. We communicated with the SAC Command Post, who patched us into the weather station. The meteorologist advised us we would break out of the clouds shortly. We continued to climb. Now we were at 20,000 feet and the aircraft would not climb any higher. Meanwhile, the B-47 bombers were directed to hold with engines running. Again the weather station was consulted. We were advised to move off the west coast of England and we broke out into the clear.

Finally, the bombers were released for takeoff. By the time we refueled them, they requested more than the 3000 pounds of fuel we were scheduled to give them. After giving them the requested offload, we diverted to Prestwick, Scotland to refuel in order to be able to recover back to Fairfield.

Active Duty in SAC KC-97G Tankers
SAC beat up their aircraft. The KC-97 was originally designed as an airliner for a gross weight of 153,000 pounds. SAC loaded them up much higher. Heavyweight takeoffs worked the engines hard. Engine life suffered. Refueling power at 15,000 feet altitude was METO, or Maximum Except Takeoff. The turbo superchargers

"crammed" air into the engine intake manifold to trick the engines into performing as they would at a lower altitude. We always feared an engine failure during refueling. In that event, the procedure was to announce over refueling radio frequency, "Breakaway, Breakaway, Breakaway." The bomber dropped off the tanker boom and slid back. The tanker crew handled the emergency at this point.

Aircraft on alert were loaded to the landing gear weight limit of 175,000 pounds. This would be the aircraft weight if it were launched to go to war. Takeoffs at this weight were dicey. Acceleration checks between 60 and 100 knots on the runway could be as much as 17 seconds.

Equipment failures were frequent. Occasionally, my radar malfunctioned. Other aircraft malfunctions were common, such as radios, engines, hydraulic systems, what have you. No two flights were the same.

Weather was a variable. Sometimes a weather front had to be penetrated. On active duty, on a Mass Gas, the formation tried to penetrate a weather front, unsuccessfully. We flew parallel to the front for quite a while, looking for an opportunity to penetrate it, using the navigator's radar. Eventually, we gave up, scrubbed the mission and returned to Otis AFB. Other times, turbulence was a problem.

Once, on the refueling track, I directed our aircraft to depart the Orbit for the ARCP. I was running the rendezvous with a B-47. Suddenly, the pilot did a wing over to the left, while stating over interphone, "Son of a Bitch!" A Royal Canadian Air Force (RCAF) C-130 descended across our flight path. Air Route Traffic Control (ARTC) was contacted but had no contact with the other aircraft. That was a close call.

Memories of my First Overwater Flight as a Navigator
I had many flights across Baffin Bay, both to and from Sondestrom AB, Greenland, for alert tours. Suddenly, my crew was assigned a mission to England. Now, I was going to make my first long range

overwater flight, across the Atlantic. My stomach dropped to the floor, and I was nervous.

There was a tanker crew from Westover AFB deadheading in the back of our aircraft. Also, there were maintenance personnel, as well as others. Someone attempted to insert a soup heating coil into a DC electrical receptacle in the main cabin of the aircraft. My Radar Feed circuit breaker popped, resulting in the loss of my radar, radar altimeter and Loran electronics. The spare navigator helped me find and reset the circuit breaker.

The rest of the flight was uneventful, initially. About halfway across the ocean, the flight engineer checked off interphone to service the engine oil supply. There was a 50 gallon drum of engine oil strapped to the floor, just behind the cockpit bulkhead, next to the single point oil filler pipe. There was a hand pump on the top of the drum, to facilitate adding oil. By adding oil through this service point, the flight engineer was able to distribute the oil to each engine requiring servicing, in flight. Sometime later, the flight engineer reported we were using too much oil in one of our engines. The offending engine was shut down as a precaution. We proceeded on three engines. When we made radio contact with St. Mawgan RAF Base, on Lands End, we declared an emergency and landed.

The RAF personnel were much more aviation minded than our own people. As soon as we parked, we were surrounded by RAF personnel who wanted to see our aircraft. We hitched a ride to the officer's club for lunch, and were seated at the base commander's table. At the last minute, the commander was called away, but we enjoyed the attention and the meal. The oil leak was fixed and we proceeded to our destination, RAF Brize Norton near Oxford, England.

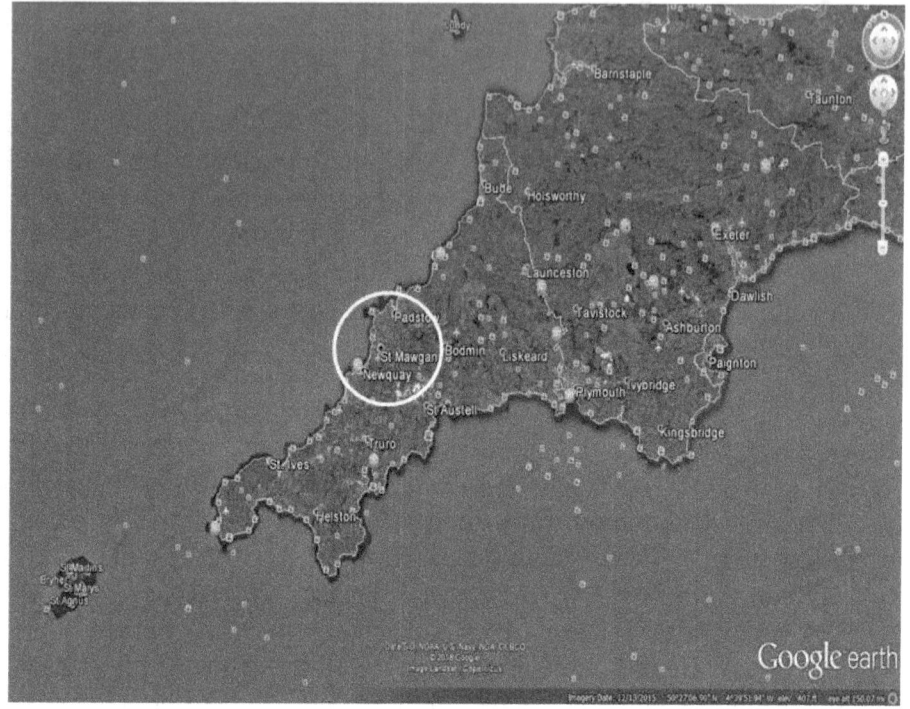

St. Mawgan RAF Base in England

Our tanker was put on alert beside the alert B-47 bombers. An English pig farmer resided just off the base active runway. He owned a small aircraft and had his own airstrip. Whenever he wanted to fly, he just started his engine and took off. He had no communication with Brize Norton RAF Base control tower. This caused considerable consternation, as the active runway had to be available in case of an Emergency War Order (EWO) launch of the bomber alert force.

B-47 Bombers
Ref: en.wilipedia.org

SAC worldwide exercise in 5 November 1964

The crew to which I was assigned completed a home alert tour (at our home base, Otis, AFB). We were entitled to compensatory time off, consisting of 50% of the alert time. We were released from alert Thursday morning, and were entitled to time off until Monday morning. There was a worldwide SAC exercise scheduled for Friday. The 19th ARS was scheduled to take off at 1500 hours (3:00 P.M.). Marilyn and I ran an errand to Hyannis and returned in time to watch the 19th ARS MITO in support of this massive exercise. The MITO was a takeoff procedure where each aircraft took off 15 seconds behind the preceding one. (This was how the alert force was to launch under wartime conditions.)

We parked at base operations just before 1500 hours and observed the six aircraft with engines running, waiting for takeoff time. Precisely at 1500 hours the lead tanker rolled down the runway. I expected number two to roll 15 seconds later, but it didn't happen until one minute later. Each of the other aircraft took off at one minute intervals. I didn't know what happened.

Later, I learned the KC-97G unit at Pease AFB, Portsmouth, New Hampshire launched their tankers two hours earlier. Number four tanker in the MITO stream crashed and blew up off the end of the runway. SAC cancelled the MITO takeoffs for the balance of the exercise, pending a formal accident investigation.

G Dornfeld posing in his thermal flight suit, prior to deploying to Sondestrom AB, Greenland for a week long alert tour.

The mission of all SAC flying units was to be prepared to go to nuclear war, if attacked pre-emptively by the Soviet Union. The theory was, by being prepared to wipe out the Soviet Union in response to an attack, there wouldn't be one. Our flying missions were intended to maintain aircrew proficiency. As an air refueling tanker unit, we practiced navigation and air refueling. This was serious stuff, and there was a lot of tension constantly.

Pilot Proficiency Missions
Interspersed with these missions, we occasionally flew pilot proficiency missions. These were flights where the two pilots accomplished their required takeoffs, landings, precision and non-precision approaches. The flights lasted approximately five and a half hours. Sometimes, when things became too boring, we just flew around for a while.

There was a nudist colony somewhere on Cape Cod. One day, while on a pilot proficiency mission, we decided to look for this facility. The copilot called the tower and told them we would be heading east along Cape Cod for a while. The tower knew what was going on, and replied, "Take a heading of 096 degrees." They were helping us out without admitting it. Unfortunately, we didn't find the nudist colony. DRAT!

Since SAC had the integral crew concept, I was along for the ride as a crewman rather than a navigator. My job was to keep track of the approaches and landings each pilot accomplished. There were precision and non-precision approaches. After each approach, the pilot flying the approach made the decision to make a "go-around," a touch and go landing or a full stop with propeller reverse thrust and taxi back. Most approaches were "Missed Approaches" as we could accomplish more approaches per hour this way.

On the Before Landing checklist, after the pilot lowered the landing gear, I had to observe the nose gear was down visually. This required unstrapping the seat belt and shoulder harness, turning the driftmeter to observe the nose gear, and reporting "Nose Gear-Down Visual." Then I had to reapply the seat belt and harness. The process became repetitive.

The exciting part was the touch and go landing. After the landing gear were firmly on the runway, the wing flaps were set to 55%, engine RPM to 2700 and maximum power applied.

Official Photograph of 2nd Lieutenant GR Dornfeld when

assigned to 19th ARS

Allotted Flying Hours

Most people are familiar with government mentality; every allotted dollar must be spent by the end of the fiscal quarter, or the government agency will not get the same amount for the next quarter. The same held for the Air Force, relative to the allotted flying time per quarter. (Each hour of flight time is costed as a dollar amount.)

We approached the end of a quarter with approximately 4-5 hours left of our allotment. All flying training requirements were complete, but some hours were left. The end date for the quarter was a Saturday, normally not a duty day. Two crews and aircraft were readied for flight on Saturday. One crew took off and flew a pilot proficiency mission, shooting landing approaches, missed approaches, touch and go's, and full stop landings, to use up the hours. My crew was on standby, waiting in case the first aircraft broke, so we could then take off and fly the remaining hours. Government Insanity!

Celestial Navigation Stressed

It is difficult to realize today, with the availability of GPS to everyone, aircraft navigators were required. In reality, flight across the continental United States did not require a navigator, as the ground radio navigation facilities and the aircraft navigation radios were sufficient for a pilot to self-navigate his aircraft. For example, airlines stopped carrying navigators years earlier. However, SAC trained for war, while hoping its readiness would keep the peace. SAC had the motto, "Peace is Our Profession."

The following is a quotation from SAC Manual 55-10, "Electronic equipment is complex, can be detected by the enemy radars, can break down, and is vulnerable to jamming and severe weather. On the other hand, celestial equipment is relatively simple, virtually foolproof, and is not susceptible to hostile action. The celestial method was considerably more involved, slower, and less accurate than electronic devises; however, with practiced procedures, the skillful navigator can guide the airplane to the target and return with nothing more than a sextant, an accurate watch, a suitable chart, a set of navigation tools, an almanac, a computer and a set of HO-249 sight reduction tables."

Accurate Watch

An accurate watch was essential for accurate celestial navigation. This was because an error of four seconds could mean as much as a one mile error while performing celestial observations. When I first reported to the 19th ARS, as a 2nd lieutenant, I checked in with the supply organization. They issued me a "used" watch with mediocre accuracy. The explanation was the used watch was all they had. However, when I took my first annual physical at the base hospital, the enlisted Air Force corpsman took my heart pulse with a new issued watch. Obviously, the corpsman had a connection with the supply sergeant I did not have. And, as a new navigator, I was not about to make waves. Apparently, the hospital corpsman needed a new watch to measure the heart pulse of personnel during their annual physical exams more than the crew navigators needed the new watch for celestial navigation. Government insanity!

Trapped in the Latrine

During a deployment to a remote site for an alert tour (called reflex), I performed a day celestial navigation leg (minimum of two hours planned duration and nine sun shots with the sextant). The sextant mount was located on the ceiling next to the navigator station, and an adjustable-height round stool was used to adjust the individual navigator's standing height so he could comfortably sight the celestial body. The stool resembled a round piano stool, and was called by that name. When the stool was in place, and locked in position, the hinged folding door to the latrine in the cockpit could not be opened.

We had on-board a female passenger who was in the latrine when it was time for me to get my last three sunline shots. (Each shot was a two minute sighting, using a mechanical averager on the sextant.) In the middle of one of my two minute shots, she attempted to come out of the latrine, but the folding door kept banging into my stool. I tried to explain to her to wait a minute, over the loud noise of the engines, but she thought I was playing a joke on her, and became vocal and uptight. When I finally was able to release her from her captivity, she blasted me. She didn't buy my explanation!

Body Heater Fire

Just before deploying to Sondestrom Air Base, Greenland, for a reflex alert tour, our crew was briefed by the squadron safety officer. He reported a KC-97 tanker from Plattsburgh AFB (Plattsburg, New York) had a body heater fire the crew thought could not be extinguished, and they bailed out. As background, the KC-97 had two large cylindrical gasoline-fed heaters providing heat for the interior of the aircraft. These two heaters were located in the lower forward compartment under the main floor. The navigator was supposed to be the primary firefighter in the case of a body heater fire. At the direction of the pilot, the navigator was to don the fire fighter's mask with oxygen bottle, take a portable fire extinguisher, and fight the fire. There was one problem; nobody had ever experienced a body heater fire.

Flash-forward to this safety meeting, we were told the Plattsburgh AFB crew felt they had not been able to extinguish the fire, due to the excessive amount of smoke, and had successfully bailed out of the aircraft. The aircraft flew, on autopilot, until finally, out of fuel, it glided to a gentle crash landing in northern Canada, in a bog. A crash investigation team visited the scene and concluded the fire had indeed been extinguished. The absence of prior body heater fire experience resulted in a lack of understanding. An extinguished fire still resulted in heavy smoke, but the fire was indeed out. This was embarrassing to the crew.

Major Stewart Preston
Major Preston, had been a bomber pilot in WWII, and was finishing out his twenty years in the Air Force, when I knew him. He was the aircraft commander (pilot) of one of the aircrews in our squadron. The story was, during WWII, his bomber had been severely disabled during a bombing mission over Germany, and the crew elected to "stray" into neutral Switzerland, and land, whereupon they were interred for the war duration, in a hotel.

He was an interesting person, as he was a body builder and had a tremendous upper body build, with no neck. The enlisted boom operator on the crew was a wisecracking streetwise youngster from the Bronx, who enjoyed constantly "razzing" the navigator, in a good-natured way. To listen to them, the uninitiated swore they were having a serious argument. The argument always ended up with the boom operator grabbing the navigator's plastic navigation plotter and snapping it in half across his knee. Then the boomer walked back to the rear of the aircraft, and got a spare plotter from his briefcase he had just for these occasions and gave it back to the nav.

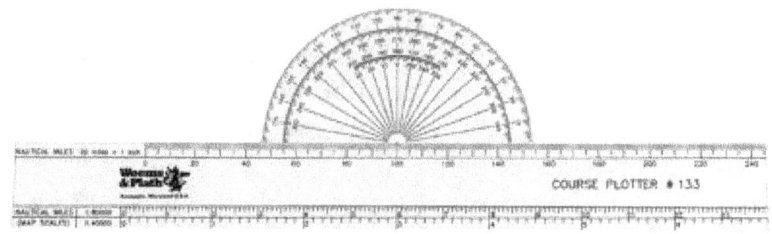

Navigator's Wheems Plotter

The copilot never had a moment to relax while flying, as he had to constantly watch Major Preston, because the he had a tendency to do things that could get the crew in trouble. Late one night while in the Orbit pattern, the navigator was talking to the bomber pilot during the rendezvous on UHF radio and gave Major Preston the compass heading to depart the Orbit for the Air Refueling Control Point. Suddenly, the navigator looked up at his N1 compass, and the aircraft continued to turn past the departure heading. He called out on interphone, "Hey, what's going on?" (Not the best interphone discipline; he should have stated over interphone, Pilot; Nav, You passed my heading. Return to a heading of 276 degrees.) The copilot looked over at Major Preston in the left seat; he had the control yoke in his hands, in a left hand turn, and he was fast asleep!

I attend reunions of the 19th ARS every two years. Nobody knows what happened to Major Preston, and he is presumed deceased. The navigator, likewise has disappeared. However, at the 2008 reunion in Las Vegas, the copilot attended, and we all had good laughs about the old times. Just recently, I received an email stating he has died of ALS. Unfortunately, veterans of the 19th ARS range in age from the mid-70s to the mid-80s, and are dying at an accelerated rate. I do not know how many years we can continue to have reunions, so I try to make them all.

Tanker Missed England
Our squadron navigator was a major who was a World War II holdover, finishing his 20 years for retirement. Although he had over 3000 hours of flying time, and therefore was a master navigator, the reality was he was such a poor navigator he could not be trusted on a crew. The squadron put him on staff, as squadron navigator where he could do no harm. He took leave and dead-headed on a Pease AFB (Portsmouth New Hampshire) KC-97 tanker to England. The navigator on this aircraft was a 2nd lieutenant on his first flight over water. The aircraft missed England completely, and the crew couldn't determine where they were. Every radio station available was tuned in and finally they determined they were coasting-in to Spain.

Our major swore vociferously and repeatedly he stayed in the back of the aircraft the whole time, but everyone in the squadron knew he must have come up and volunteered to "take the seat" for a while, relieving the inexperienced navigator. Any one of us would have done the same thing, had we been in that position. The poor 2nd lieutenant met a Flying Evaluation Board, and I believe they stripped him of his wings.

Strategic Air Command (SAC) Facilities Inspection
While we are on the subject of our squadron navigator, we were about to have a SAC Facilities Inspection. Everything in our operations building was cleaned up and made neat. The building had a central corridor with offices leading off both sides. Our major kept peeking out the navigator office into the corridor to see when the team came through the entry double doors. Finally, the doors flung open, and he whispered in a voice heard a mile away, "Here they come, fellas." The team made their way from one office to another, until they came into the navigator office. The general said to the major, "By the way, major, it is 'TENHUT,' not 'here they come

fellas."

While I am on the subject of the SAC Facility Inspection, the alert facility briefing room had a closed circuit television hooked-up to the base weather station. Each morning, during the daily briefing in case we went to war that day, we had our weather briefing. Someone had used a grease pencil to depict the "ban the bomb" symbol on the front of the TV screen. The peace movement was active in the early 1960s and used the "ban the bomb" as its rallying symbol. The first day, it was a joke. After that, nobody noticed the symbol; it was there, but nobody noticed it anymore. During the SAC Facility Inspection, one of the inspectors was a bomber general, and he took off on the symbol. Boy was he steamed!

Ban the Bomb Symbol

Cuban Missile Crisis

Our crew was assigned a weather scout mission on October 22, 1962. There was to be a deployment back to the U.S. of three B-47s from a reflex alert tour in Europe the following day, and the weather in the Atlantic Ocean needed to be verified.

We took off from Otis AFB at 1455 hours local time (2:55 PM), flew to Bermuda and returned to Otis AFB, while recording the weather every hour. On the return leg of the flight, in the early evening, we were challenged with our on-board classified verification documents, at each position report. We couldn't understand what was happening until after we landed at 2345 hours local time (11:45 P.M.), and found out President Kennedy had been on television earlier in the evening and had announced evidence the USSR had put nuclear-tipped missiles in Cuba. Mother called me and asked me what was going on. I replied the Air Force doesn't tell us anything, and all I knew was what was on the television news, which was the truth.

Immediately, all aircraft and crews were generated to a war footing. Our alert force of six aircraft and crews were in Sondestrom, Greenland, but all the rest of our force went to a 30-minute posture; we had to be next to the telephone at our quarters and be able to drive to the squadron operations building within 30 minutes, in flight suit, ready for deployment.

All the rest of our tankers on the ramp were upgraded to alert status, ready to go to war. The classified war plan was on board, and the SAC security police patrolled all around them, day and night, with limited access points, where aircrews and maintenance personnel could pass through. These aircraft were to be the second wave of the retaliatory strike force.

Several aircraft and crews from various squadrons in the U.S. were ordered to Goose Bay Air Base in Canada to be part of a special tanker task force. The objective was to have the tankers as far forward along the bomber route as possible to top off their fuel tanks.

I read in the Boston newspaper, B-47s from Pease AFB in Portsmouth, New Hampshire were redeployed to Logan airport in Boston, to disperse their strike force. There was absolutely no flying for training during this period of time. This was serious business.

After about three weeks, the crews in Greenland had to be relieved. On November 16, 1962, a C-54 crewed with higher headquarters staff from Westover AFB flew to Otis AFB to pick up six crews (including ours) to rotate them to Sondestrom AB, and relieve the isolated crews who had been stuck there during the crisis. Those were quite heady times. Much later, we were able to look back and realize how close the U.S. came to nuclear war with the Soviet Union.

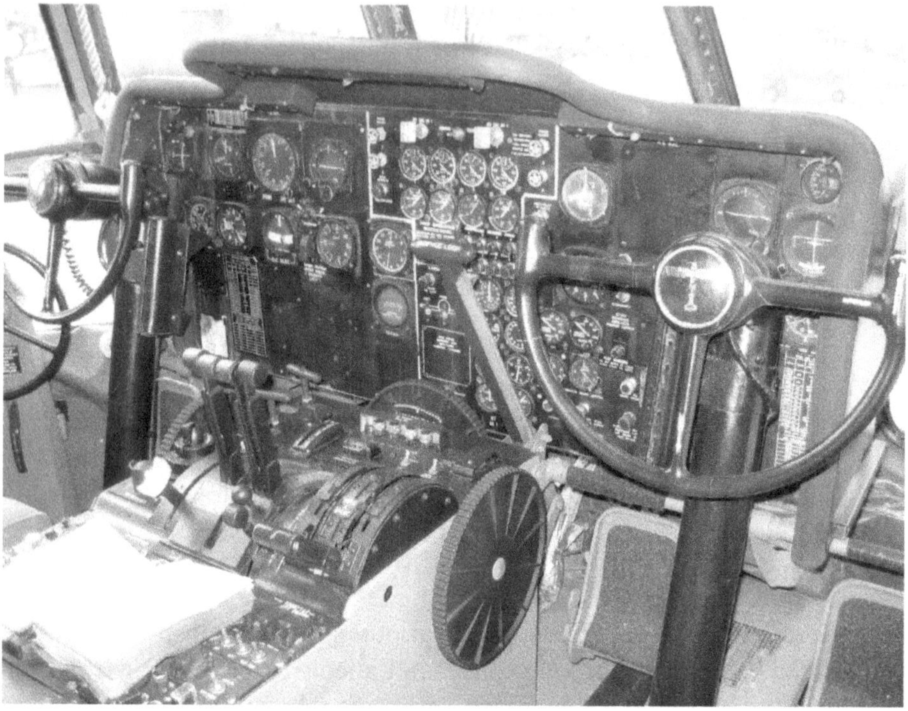

Pilot station is on left, Copilot station is on right.
Photo by Author

C-54 Similar to staff aircraft that rotated aircrews to/from Sondestrom AB, Greenland
Ref: amcmuseum.org

SAC KC-97G Air Refueling Tanker
Ref: commons.wikimedia.org

B-47 Bomber
Ref: en.wikipedia.org

Favor Buying Set of Golf Clubs

Word spread throughout the squadron we were going to England.

Engineer, Set Refueling Power

The day before we left, there was a knock on the door of our government quarters on Otis AFB. It was our flight surgeon from the base hospital. He heard we were going to England, and requested I buy him a set of golf clubs at the base exchange there.

He gave me a manufacturer priority list; first choice was Macgregor, second was Dunlop, and third was Wilson. I agreed and he paid me $25 in cash. (It was a pretty hefty sum back then, in 1964.) When I visited the large Army and Air Force Exchange in London, at Ruislip, it had only the Wilson brand in stock, and, as it was on the list, I bought the Wilson set of golf clubs. I then had to babysit the golf clubs whenever I had free time, and drag them back with me to Otis AFB.

I called the flight surgeon upon arrival, and he picked up the goods at my quarters. He was steamed at me for getting him his 3rd choice, as he really wanted Macgregor! Wow, some "chutzpah." I have never seen this doctor at squadron reunions, and I wonder if he remembers this incident.

Flight Suits in the Officer's Club

Late in my tour of duty in SAC, our unit was directed to release one of our KC-97 tankers to the Air National Guard (ANG), and the directive was by aircraft tail number, #340. The director of maintenance was not about to give up this particular aircraft with the relatively low-time engines, so he had the engines swapped-out with higher-time engines. Then we had to fly one last mission on March 12, 1964, with this aircraft, before releasing it to the ANG. It was a two-ship formation refueling, with our ship as number two.

Everything went according to the book, until the end of refueling of a B-47 bomber. After fuel transfer, we cleared the air-refueling track, and the pilot directed the flight engineer to pull back the engine power from "Refueling Power" to "Cruise Power." Just as this was done, the number two engine (inboard engine, left hand side) backfired badly, and the pilot directed the flight engineer to shut down the offending engine.

Engineer, Set Refueling Power

The "book" says when an engine is shutdown in-flight, the pilot is to land the aircraft at the nearest suitable airfield. After the engine was shut-down, the pilot asked me for the nearest suitable airfield. I replied it was Plattsburgh AFB, New York. Our pilot called the tanker leader on UHF radio frequency, and stated we had shut-down number two engine and were diverting to Plattsburgh AFB.

Lead called back and asked "Are you aware Plattsburgh AFB no longer had KC-97 aircraft assigned, and therefore no longer had KC-97 maintenance?"

Our pilot, replied, "Are you aware we are flying on only three engines?"

We diverted our aircraft into Plattsburgh AFB, just in time for lunch. After bumming a ride to the officer's club we attempted to be seated for lunch, but were rebuffed, as we were in flight suits. In those days, flight suits were not allowed in the officer's club. We argued and argued, and finally officer's club officials relented, but only after putting a folding screen around us in a corner, so nobody could see us. The Air Force of the day really didn't have its priorities straight!

My Introduction to Firing a Flare

During a night rendezvous with a B-47, after the pilot made initial radio contact, he turned over the UHF radio to me to conduct the rendezvous. On this particular night, there was limited visibility in the refueling area, and the B-47 pilot informed me on UHF radio frequency his radar was inoperative, but he wanted to get the refueling accomplished. With the bomber radar out, the backup rendezvous system was the APN-12/76. My APN-12 rarely had a signal. I was told the B-47 radar navigator needed to unstrap from his ejection seat to rotate and turn on his APN-76. (I do not know if this is the case.)

All of a sudden, a signal appeared on my APN-12 scope, and I was able to vector the B-47 in behind us. The bomber pilot requested a flare to aid in seeing the tanker. There was a pouch filled with flares next to my seat, and a "VERY" flare pistol mounted on the side of the aircraft. The aircraft had a double skin, so the theory was to insert the pistol into its mount on the inner skin of the aircraft, pull the handle to open the outer skin, load a flare in the pistol, and fire the flare pistol. The navigator was taught how to do this, in theory, but never allowed to "waste" a flare to practice.

Here I was, needing to actually fire a flare for the first time. Now I was a city boy, and never was around guns, so I was not familiar with opening the gun breach and inserting the flare. I took the gun out of the mount, and installed the flare backwards, reinserted the flare, opened the outer skin, and pulled the trigger. Of course, the trigger just went "click" and the flare just got sucked out when the door was opened.

The B-47 pilot stated he could not see the flare, and I repeated the process several times with the same result. The flight engineer looked over, saw the problem, and said, diplomatically, "Let me help you with that, sir." After successfully firing the flare, the B-47 pilot confirmed the visual sighting. I learned how to fire a flare gun the hard way! Thank G-d the flares I "shot" didn't hit the number 2 propeller.

Occasionally, the lack of aircraft availability required we perform a "hot turnaround." After an aircraft completed a refueling mission and landed, the crew changed out with a fresh one. The aircraft was parked with engines running, and the fresh crew bucked the propeller slipstream and entered the lower aft entry door. Then the crew from the first mission exited the same way. Usually, the cockpit was hot and sweaty. However, after takeoff and during climb, the interior temperature became more comfortable.

Hurricane Ginny Evacuation
On 29 October 1963, Hurricane Ginny "marched" up the east coast. I kept an eye on the storm progress on the nightly news. This evening,

as I was preparing to go to bed, the late evening news indicated we were about to be hit by the storm. I couldn't understand why we didn't evacuate our aircraft. Just as I was falling asleep, the telephone rang. It was squadron operations. They told me to get down to the operations building immediately; they were evacuating the aircraft. When we reported, we were handed the flight plan already made out, to fly to Plattsburgh AFB in New York.

We reported to our assigned aircraft, started engines and waited for our clearance to takeoff. Of course, everybody on the east coast was evacuating; military, commercial, business and private aircraft. The air traffic control system was saturated.

We rode out the hurricane with engines running, in our parking spot, pointed into the wind. Finally, we were granted clearance to takeoff at 0445 hours. By this time the hurricane had substantially past us by.

A friend of mine was on a crew on alert at Otis AFB at the time. The alert force of six tankers was downgraded at the last moment, and directed to takeoff immediately. Unfortunately, there was insufficient time to reduce the weight of the aircraft. All six aircraft took off at 175,000 pounds gross weight. I was told two of the six lost an engine on takeoff. This event must have been exciting.

Search Mission
After coming off an alert tour, Marilyn and I visited family north of Boston. I received a recall telephone call. Even though my crew was on "compensatory time off," our crew was needed as backup. A flight of two B-47 bombers was returning from a reflex alert tour in Spain to MacDill AFB, in Tampa, Florida and had disappeared. A search mission with our KC-97 aircraft was organized to look for survivors in the Atlantic Ocean. Although I was not actually launched on this mission, our crew was backup. Three aircraft were launched, flew search patterns for 12 hours, and recovered into Kindley Air Force Base, Bermuda for crew rest. The crews launched so fast, they did not have time to pack civilian clothes, and

encountered the "no flight suit in the officer's club" rule. More Air Force insanity!

After crew rest the aircraft flew another 12-hour day of search patterns. Unfortunately, remains of the two aircraft and crews were not found. The good news was there were no flight crews in sweat-stained, grubby flight suits in the officer's club!

First Flight after an Alert Tour
The aircraft were rotated off alert status on a different schedule from aircrews. Once off-alert, the fuel was downloaded to a reasonable weight and the aircraft was put into the schedule for training missions. The first flight after an alert tour was scored, as this would have been the flight should the aircraft be launched to go to war from alert status.

KC-97F refueling a B-47 bomber
Ref: commons.wikipedia.org

"Mass Gas" or Cell Missions

The stress we were all under took its toll. We didn't want to screw up and let our peers down. The "Cell" mission was the most stressful. A cell was a formation refueling mission. Typically, a cell consisted of six tankers refueling three B-47 bombers. The lead tanker accomplished the navigation, rendezvous and the serious navigation home.

The takeoff was at one minute intervals with airborne radar used to form up and maintain one mile in trail, stacked up at 500 foot vertical separation. The leader accomplished the navigation and rendezvous with the bomber cell. Once we departed the Orbit pattern to rendezvous, the tankers echeloned 45 degrees to the right and flattened out to the same altitude. Each bomber refueled from one tanker, then slid over to hit the second. It was difficult to accomplish a serious navigation leg on the way home, because the aircraft headings were not under the control of the navigator, but rather whatever was necessary to stay with the cell leader. There was no relaxing, because a mission abort by the lead tanker resulted in the next aircraft assuming lead responsibility.

Sometimes we refueled a B-52 bomber
Photo by a peer navigator

137

Engineer, Set Refueling Power

Deactivation of 19th Air Refueling Squadron

While I was on active duty in the Air Force, it never occurred to me the aircraft in my squadron would be retired. Silly me, I just felt operations were to go on forever.

When we pulled alert tours, a week at a time, the policy was to grant the aircrews compensatory time off, representing 50% of the time on alert. Aircrew changeover day was Thursday morning, so we were off-duty until Monday morning.

In the fall of 1965, after an alert tour, Marilyn and I returned from a shopping trip to Hyannis. As was my custom, I entered the base back-gate. This ran around a perimeter road past the alert facility. As I rounded the bend, I noted there were no tankers on alert! I found out the squadron had been notified it was to be disbanded, and the alert force was downgraded. I had previously requested a release from active duty, so there were no consequences for me. My crew was ordered to deliver a tanker to the Air Force "boneyard" at Davis Monthan AFB, Tucson, Arizona. Since this was to be my last flight on active duty, I bowed out when another navigator volunteered for the trip.

SAC Alert

All of the worldwide-deployed SAC bombers and tankers had 50% of aircraft and crews on 24 hour alert in hardened alert facilities, collocated next to "Christmas Tree" alert aircraft parking stubs.

This parking area for the alert force aircraft, when viewed from above, looked like a Christmas Tree, with each branch a parking location for an alert aircraft (see photo on page 110). In the event of a war launch, this allowed any aircraft ready to taxi to move toward the active runway for an immediate takeoff, without being blocked by an aircraft not ready to taxi or with a major mechanical problem. The alert facility was referred to as the "Mole Hole," as there were no windows.

Engineer, Set Refueling Power

We just finished refueling a B-47 and the bomber initiated a climb back up to altitude
Photo by Author

SAC assigned aircrew members as an integrated crew. This meant the crew trained together, flew training missions together, and pulled alert tours together. Under these circumstances, it was difficult to maintain a proper distant relationship between officer (pilot, copilot, and navigator) and enlisted crewmembers (flight engineer, and boom operator). Life for the alert aircrews was somewhat boring. With the exception of the first day, the crews were pretty much on their own. There were recreational facilities; pool tables, Ping-Pong tables, television lounge, library, gym, and a cafeteria where three meals a day were served, and a movie at night. The major in charge of the alert facility did his best, but sometimes the movie was of an old vintage. When the crews complained, the major reminded us about how much better we were than the Polaris submarine crews who were also on alert, but for months at a time.

All of the facilities were on the ground level. The sleeping rooms were underground. The pilot on the crew had his own room, the navigator and copilot shared a room, and the flight engineer and boom operator shared a room. Each day after breakfast, there was an aircrew briefing to obtain the weather forecast for the launch base and the classified routing for the war mission. Then the crews went out to their assigned aircraft and performed the "Cocked" preflight, to ensure the aircraft was truly ready to go to war. Maintenance "squawks" were fixed at this time. When alert was pulled at the home base, there were unlimited visitation rights with family in the parking lot, just outside the secure gate to the facility. Many a child was conceived in the back of a station wagon in the alert facility parking lot.

Reflex

Alert tours were not limited to the home base. A squadron could be assigned an alert posture at a remote site to further the mission (an operation referred to as Reflex.) For example, the 19th ARS was assigned reflex alert at Sondestrom Air Base in Greenland for one year. My crew was among the first to deploy to Sondestrom AB on August 21, 1962. Crews flew to Sondestrom every week to relieve those on alert. Three aircraft flew up, each with an extra crew as passengers. Then six crews relieved six crews on alert, and three aircraft flew home, again with an extra crew in each.

Engineer, Set Refueling Power

Aircraft Based at Otis AFB

McDonnell F-101B

Ref: http://en.wikipedia.org/wiki/File:F-101A_Voodoo.jpg

Lockheed EC-121

Ref; http://en.wikipedia.org/wiki/File:Lockheed_EC-121D_Thailand_1972.jpg

Boeing KC-97G

Photo by Author

Sondestrom Air Base presented some challenges. First, it was of sufficiently north latitude for 24-hour sunlight in summer and perpetual night in winter. The air base was located on the west coast of Greenland, inland, at the end of a 90 mile fiord. The runway grade enabled takeoff and landings only be made downhill toward the open sea. This was because there was a small hill at the inland end of the runway the fully loaded KC-97 aircraft could not clear on takeoff. Once a week, a Scandinavian Airlines System (SAS) DC-8 jet commercial airliner landed enroute from Los Angeles to Copenhagen, to refuel, and this aircraft had no problem negotiating the hill on takeoff!

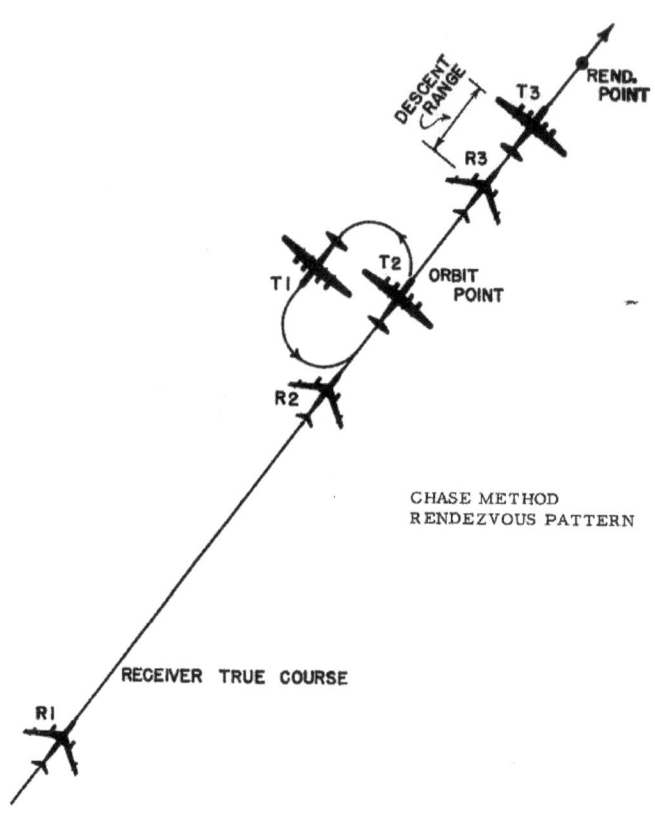

REND.
POINT

T3

DESCENT
RANGE

R3

T2 ORBIT
T1 POINT

R2

CHASE METHOD
RENDEZVOUS PATTERN

RECEIVER TRUE COURSE

R1

Engineer, Set Refueling Power

Overview of Typical Air Refueling Track
Air Force Air Training Command Course
Operational Techniques and Equipment
Provided by Author

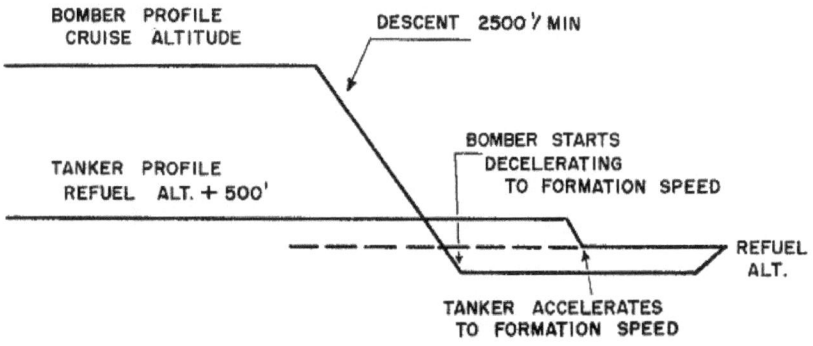

Side View of Typical Air Refueling Track
Air Force Air Training Command Course
Operational Techniques and Equipment

Sondestrom AB, Greenland

On the first deployment, our aircraft carried food supplies for the Sondestrom AB Officer's Club, from a contracted source in western Massachusetts. The officer's club manager, a young lieutenant, gave an open bar to the primary crew of the aircraft delivering the supplies. I drank very little, but the other officer crewmembers soaked up the booze. Pretty soon, there were four or five drinks piled up in front of me. The club officer was steamed at me for not drinking everything in front of me. Heck, I didn't order the drinks, they just kept coming!

Engineer, Set Refueling Power

Photo by Author
Sondestrom AB in Winter; Sun did not rise above Horizon before Setting

Photo by Author
1st Lieutenant Dornfeld Posing in front of KC-97G just after arrival at Sondestrom AB

Photo by Author
19th ARS KC-97G Tanker just Arrived and was Parked on the ramp at Sondestrom AB

Photo by Author
Two KC-97G aircraft from the 19th Air Refueling Squadron parked on the ramp at Sondestrom AB, Greenland, prior to being put on alert. Note the "Small Hill" in the Background, at the end of the Runway, just to the left of the control tower.

Sondestrom AB, Greenland
Photos by Author

When we left the club to walk to our quarters, it was approximately 2300 local time. The sun shined brightly in the sky, somewhat blinding our eyes after coming out of the darkened bar of the club. We were completely unprepared for continuous sunlight in the summer, 24 hours a day. It just didn't compute! As I indicated, our first deployment was flown in the daytime. Shortly afterward, it was determined we needed to fly to Sondestrom at night, so the aircraft arrived in daylight. The maintenance people would then have all day to work on the aircraft, if necessary, to correct faults to make the aircraft ready for alert on changeover day. So, we ended up flying all night on future deployments.

On December 19, 1962, my crew deadheaded in the back of an aircraft. I had the beginnings of a cold. The primary crew pilot liked to manhandle the aircraft. As he called for the "Descent" checklist, he pointed the aircraft nose down and made what I can describe as a "Jet Penetration," which was a steep dive to lose altitude quickly. I couldn't keep up clearing my ears with the rapid change in cabin altitude, and ended up with an ear block. I was taken to the flight surgeon after landing, and he put a tube in my nose with a bulb on the end of it. He told me to say "Chocolate," and at precisely the right instant, he squeezed the bulb and my ears popped, and I was fine.

Marilyn and I lived in base quarters at Otis AFB, which could be described, today, as a two-bedroom townhouse. There was a small hole in the basement concrete, and at this time of year we got crickets in the basement. The noise was troublesome.

Two KC-97G Tankers from the 19th Air Refueling Squadron on the Ramp as Sondestrom AB, Greenland. Note the hill to the left of the control tower.
Photo by Author

Sondestrom AB in winter. Sun actually didn't rise. It got light for a while, and then the sun set.
Photo by Author

Engineer, Set Refueling Power

Deployment over Canada. Photo by Author

Just before a deployment to Sondestrom, I was determined to kill the crickets. I bought bug killer in a spray can, and I unloaded the whole can in the basement. Later, we took off and as I shot the sextant, my eye became itchy. By the time we landed at Sondestrom, my eye was swollen. Again, I was taken to the flight surgeon, and I was grounded for the week. The diagnosis was I was allergic to the sextant! The mission for our aircraft was scrubbed from the alert force. When our relief came a week later, an extra navigator came to substitute for me, as they would not allow me to perform my duties on the crew while flying home. I had actually developed an allergic reaction to the excessive bug killer application (overdose).

The primary mission of Sondestrom AB was to provide support for the Dye radar site on the Greenland icecap. Weekly, a C-130 aircraft from Dyess AFB, Abilene, Texas landed with skis attached to the landing gear. The aircraft landed on the wheeled gear, but the skis were for landing on the ice cap. This aircraft picked up supplies for the radar site and delivered same using the skis.

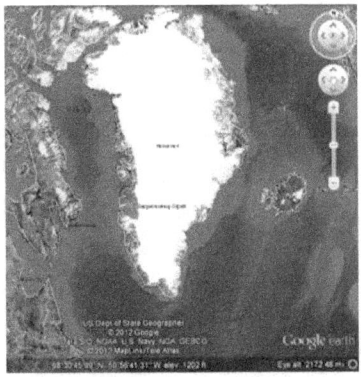 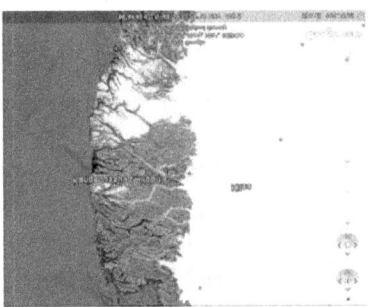

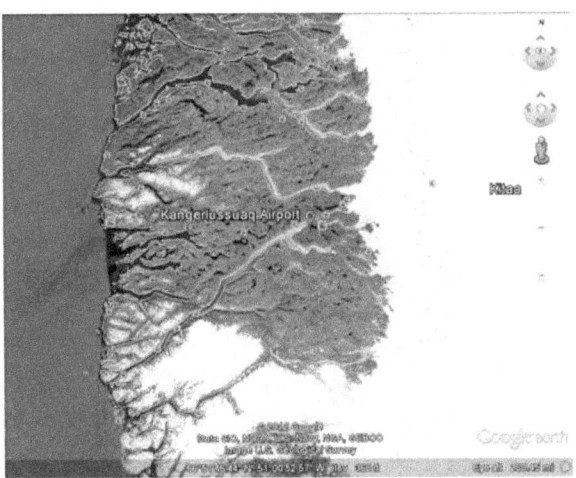

Sondestrom AB is now called Kangerlussuaq Airport

Ballistic Missile Warning System (BEMEWS)

The Air Force had completed a system of long range radar sites to provide warning of Soviet ballistic missiles coming over the polar region. This system was called BEMEWS and consisted of three radar sites; Site 1 at Thule Air Base, Greenland, Site two at Clear Air Force Station in Alaska and Site Three at RAF Flyindales, United Kingdom. The system provided a minimum of 15 minutes warning so the SAC alert force could be launched to retaliate.

An interesting incident occurred before I joined the squadron. On October 5, 1960, only five days after the BEMEWS system had become operational, at 1617 hours, the radar data on a display board in the war room at SAC Headquarters began flashing alarm level indicators. The indicators rose to level five, meaning 99.9% certainty of a missile attack! Other displays showed size and strength and possible targets of the attack. All over the world, the klaxon sounded in SAC alert facilities. At the 19th ARS, the crews responded, as a normal alert would require.

However, after engine start, when the crew received the radio authentication message in code, it was a valid launch message! As required, both the copilots and navigators validated the launch message, using the classified material on board. The aircraft taxied out for immediate takeoff, but were held on the runway, until things were straightened out. **The new radar system had picked up the moon coming up over the horizon in Norway, 239,000 miles away, and was projecting impact targets.** Luckily, it did not take too long to figure out what was happening. The radar was later reprogrammed to reject radar returns from the moon!

Early during my service in SAC, the tankers were painted orange on the nose and tail, presumably to aid in acquiring them visually. Later the paint was removed to save weight (300 lbs. per aircraft was the number quoted), and apparently because someone determined they were not particularly effective in increasing visibility.

In the winter, the "Cocked" aircraft required portable hot air heaters powered 24 hours a day, with ducts routing hot air into each engine intake, as well as into the cockpit. This was to insure a rapid start could be accomplished. The hot air ducted into the engine nacelles warmed the engine oil, resulting in reduced oil viscosity (enabling the oil to flow more freely), so the engine starter motor wouldn't work as hard. An engine start under extremely cold temperatures would be difficult, due to the congealed engine oil. Also, warming the gyros in the cockpit instruments (directional gyros, and compass system) enabled them to spin up rapidly. These heaters were necessary to facilitate a war launch in the winter. Obviously, there was no family visitation or telephone contact at this remote site.

Our crew was scheduled to return from a reflex alert tour at Sondestrom AB, but the aircraft had just had an engine change, and needed a functional flight check before it could be released to fly home. This check consisted of a takeoff, a flight around the landing pattern and a full-stop landing. It was winter and the runway was slick with hard packed snow and ice.

The pilot released the brakes, and the aircraft started to move. The boom operator was stationed at his position for takeoff, which was in the rear of the aircraft, on the left hand side, scanning the engines on that side. He reported on interphone, "Sir, the left main gear are not turning," and repeated this call throughout the takeoff. There was little friction on the runway, and the left main landing gear never did turn. It was basically a controlled skid to takeoff. Landing was a bit hairy also, with propeller reverse thrust used to slow the aircraft to the point the brakes could be used very gently. After that successful functional check flight to clear the engine fault, we were released to fly home. It was the way it was.

On redeployment from Sondestrom AB to Otis AFB, we took off and climbed to 16,000 ft. altitude. We coasted out with Greenland behind us, enroute across the Baffin Sea toward Canada. I unslaved the aircraft compass from magnetic sensing to gyro mode, so as to navigate the aircraft using "Grid" Navigation.

For my first position, all I had to work with was a single sunline using my sextant and a pressure pattern line of position. These two lines were significantly removed from my deduced position using preflight winds, I was concerned. A half hour later, again all I had were the same navigation aids, and they indicated we were drifting to the left of course, significantly. I gave the pilot an alter heading to the right of approximately 15 degrees, to hedge my bet, but I was still not sure if I was off course or not. Then I had a panic attack…did I give the pilot the magnetic heading instead of the grid heading? A quick check of the flight plan confirmed the two numbers were within two degrees of each other in this location. We were at least 40 miles off course to the left, if the sunlines and pressure pattern LOPs were to be believed.

Now, I had to make good the coast-in point, to Canada, within 10 miles and 5 minutes, and there were two other aircraft, one 30 minutes behind me, and the other one hour behind. I didn't want to be embarrassed by screwing up. Finally, I had to admit to the pilot my navigation aids were unreliable and he should radio this information to Air Route Traffic Control (ARTC at Moncton Center). In those days, ARTC did not have radar coverage in this area, and they relied on the aircraft position reports. Let the chips fall where they may. Just as soon as we sent the radio call, both tankers behind us heard the call and did the same thing. The other tankers were having the same problem. I felt vindicated. The forecast winds were inaccurate!

In the winter, the crew vehicles had an electrical heated dipstick to warm the engine oil while parked, thereby insuring the engine would start under extreme cold conditions. The procedure for aircraft engine start from the "Cocked" configuration was discussed earlier, and is repeated here for clarity, as follows: pilot, co-pilot and flight engineer ran up the lower forward entry door to their crew positions to start engines, while the navigator stood fireguard outside, with a two wheeled fire extinguisher bottle, under the wing on the aircraft right side, for the #3 engine start (engine start sequence was #3, 4, 2, and then 1), then moved behind #4 engine for its start. After both engines on the right hand side were running, the navigator bucked the propeller slipstream, ran behind the running engines, and pulled the right main wheel chocks. Then he ran behind the aircraft, and entered it through the lower aft entry door. After engines # 3 and 4 were started, the boom operator performed the same functions on the left hand side for engines # 1 and 2.

The SAC command post broadcasted a message by radio using the classified verification method (which will not be discussed here, for obvious reasons) calling for either a "Bravo" or a "Cocoa" alert. A "Bravo" alert represents an engine start only, after which the copilot called-in via radio the aircraft was ready to taxi. This time to start engines was scored.

A "Cocoa" alert required the aircraft start engines and actually taxi to the active runway, and the time was scored as each aircraft crossed the active runway threshold. This represented takeoff time. All aircraft had to cross the active runway threshold within 15 minutes of the alert. This represented the alert time the BMEWS radar network provided before enemy ICBMs could impact, thereby destroying the SAC retaliatory force.

The aircraft taxied down the active runway and exited at the nearest turnoff and returned to their christmas tree stubs, Maintenance was notified by radio to correct aircraft faults, and fuel trucks serviced the aircraft to bring the fuel tanks back up to their maximum gross weight limit. This was happening simultaneously all over the world, where B-47, B-52, and B-58 bombers and KC-97 and KC-135 tankers were on alert.

SAC Alert Force MITO (Minimum Interval Takeoff) Takeoff Procedures

Once per year each aircrew was required to demonstrate this procedure, but not at an aircraft weight of 175,000 lbs. My last year in SAC, the requirement was changed to more nearly reflect a war launch, at the heavy weight.

The new requirement was to perform the MITO at an aircraft weight of 175,000 lbs. gross weight. The KC-97 aircraft was not designed originally to takeoff at 175,000 lbs. aircraft gross weight. Loss of an engine during takeoff under these conditions could be dangerous. The procedure for engine failure in a MITO stream was for the copilot to announce in the blind, on the UHF radio "Guard" frequency (243.0 MHZ), "ABORT, ABORT, ABORT." Upon hearing this announcement, if an aircraft in the stream had not attained flying speed, the pilot was required to abort the takeoff. All aircraft at flying speed continued the takeoff. One method of lightening the weight of a tanker with engine trouble was to dump fuel through the refueling boom. This was quite an effective procedure to lighten the aircraft weight rapidly.

SAC rotated the alert aircraft every three weeks or so, to make them available for training missions and to exercise their systems. Nothing made an aircraft malfunction more than sitting for long periods of time, especially at heavy weight. Hydraulic seals leaked and electronics malfunctioned. The first flight after an alert tour was deemed a "FSAGA," or First Sortee after Ground Alert." This flight represented how the aircraft in question would have performed, had an actual launch been ordered, and the event was scored.

Air Force Minimum Requirements for a Rated Navigator
The Air Force had minimum requirements, specified in AF Regulation 60-1, all navigators had to fulfill. A flight was required every 60 days, eight cross country flights were required per year, each cross country consisted of a minimum of two hours, and a maximum of 90 hours could be flown in any 30 consecutive days. Air Force major commands (such as SAC) usually added their own requirements to complement the Air Force 60-1 minimum requirements.

SAC Aircrew Navigator Training
When not on alert, aircrews maintained proficiency using training missions. Navigation training legs were always accomplished after the refueling. These requirements were published in SAC Manual 50-4. Every six months each Navigator was required to fly (I have

forgotten how many) the following types of two-hour minimum navigation legs:

Day celestial navigation leg

-Minimum duration two hours per flight plan

-Nine celestial Lines of Position (LOP); minimum of three per hour

-In any combination of MPPs (Most Probable Position) and/or day fixes

(If only 1 LOP is available, the navigator must use his judgment to place the position between the ded reckoning position and the LOP. This position is called a MPP or Most Probable Position.)

-Two pressure pattern LOPs (reading over land unreliable, but necessary for training)

-Two celestial true heading checks

Day celestial grid navigation leg

-Minimum duration two hours per flight plan

-Nine celestial lines of position; minimum of three per hour

-Two celestial heading checks

-Two celestial grid winds

-Must steer by unslaved directional gyro

-Gyro precession determined by celestial heading checks

-All celestial observations and plotting referenced to grid north

Night celestial navigation leg

-Minimum duration two hours per flight plan

-Nine celestial lines of position; minimum of three per hour

-Two celestial winds

-Two celestial true heading checks

-Maintain 80% of scored positions within 25 NM of planned track

Night celestial grid navigation leg

-Minimum duration two hours per flight plan

-Nine celestial lines of position; minimum of three per hour

-Two celestial heading checks

-Two celestial grid winds

-Maintain 80% of scored positions within 25 NM of planned track

-Must steer by unslaved directional gyro

-Gyro precession determined by celestial heading checks

-All celestial observations and plotting referenced to grid north

Radar navigation leg

-Minimum duration two hours per flight plan

-Six radar fixes; time between fixes not to exceed 30 minutes

-Two track and ground speed winds

-Control times at terminal point

Reflex navigation leg (navigation during reflex deployment)

-Minimum duration 4/6 and/or 8 hours or more

-Above 60 degrees north or south steered by unslaved

 directional gyro

-Four hour leg minimum; 12 celestial LOPs

-Six hour leg minimum; 18 celestial LOPs

-Eight hours minimum; 24 celestial LOPs

-Above 60 degrees north/south, all LOPs plotted

 referenced to grid north

-Use planets/moon to maximum extent in daytime reflex Nav mission

Airborne Radar Approach

The navigator, as an integrated crewmember, had to monitor all of the pilot and copilot landing approaches. However, one landing approach the navigator had to make on his own, periodically, was an airborne precision radar approach. Using airborne radar alone, the navigator had to align the aircraft with the active runway. Unfortunately, the APS-42 search radar either wasn't precision enough, or this navigator just wasn't proficient enough, but I never had a really successful airborne radar approach to Otis AFB. I aligned the aircraft as best I could, and on final approach, I stood up and looked over the pilot's shoulder, and we were never aligned with the runway.

Mission Planning

The missions were posted in the operations building one week in advance. The schedule reflected the crew number assigned (Each aircrew was assigned a number; mine was T-78), the takeoff time, the Air Refueling Control Time (ARCT), the fuel offload and the air-refueling track. In the northeast U.S., the two refueling tracks used were "Tea Shop" and "Fighting Fox." Tea Shop ran East – West across upstate New York, whereas Fighting Fox ran Northeast – Southwest across the rural mountainous part of Maine.

I much preferred Tea Shop because on Fighting Fox the winds at altitude over the mountains of Maine were unpredictable, and forecast winds were rarely accurate. Additionally, the Finger Lakes across our flight path in upstate New York made excellent radar targets to fix the aircraft position.

The day before the mission, the aircrew reported to the squadron operations building for mission planning. The navigator drew up the route on the charts, and measured and recorded the route on the flight plan side of an Air Force log, the following navigation check points:

Departure Point

Climb Route

Navigation Check Points to the Air Refueling Area

Bomber IP (Initial Point)

Orbit Point and Orbit Pattern (the point that the tanker would orbit while contacting the bomber and conducting the rendezvous)

Air Refueling Control Point (ARCP)

End Air Refueling Point (EAR)

Navigation Leg Turning Points (A two hour navigation leg was generally planned after refueling on the way back to the departure base.)

Planned Descent Routing

Arrival Point

Alternate (Emergency) Airfields

Warning and Caution Areas to be avoided

Fighter Interceptor Base Climb Corridors to be avoided

Preflight

The day of the mission, the crew report time was one and a half hours before takeoff time (two hours for a formation refueling or "Mass Gas"). The pilot and navigator reported to the base operations building for a weather briefing and to obtain the forecast winds. The navigator applied the forecast winds to the planned course routing to obtain the compass headings, estimated timing between checkpoints and Estimated Times of Arrival (ETAs). A planned 15-minute Orbit time was built into the mission planning.

The pilot then filed a flight plan and checked the NOTAMS (Notice to Airmen,) posted on a board for all facilities in the United States. For example, a NOTAM indicated a particular radio facility was inoperable, or there was scheduled parachute jumping or fireworks in a particular area. Meanwhile, the copilot obtained the classified material to be carried on board, the weapon he was required to carry because of the classified, and the flight lunches (somebody had to do it.)

Mission Briefing

The pilot and navigator then joined the rest of the crew for the mission briefing in the squadron operations building. This consisted of a formal briefing on the mission, the routing, the intended receiver aircraft type (bomber), the ARCT (time the rendezvous was to be completed), the tactical call sign of both the tanker and the bomber, the mode two code settings for the rendezvous equipment, the scheduled fuel offload, and the UHF radio frequency for radio contact with the bomber. Then the squadron safety officer gave a briefing on anything of interest to the crew concerning the mission, procedures or aircraft systems.

Preflight

After the briefing, the crew picked up their Personal Equipment (PE) consisting of the oxygen mask and parachute harness. A "Bread Truck" transported the crew out to the assigned aircraft.

Upon arrival, the pilot held a formal briefing concerning emergency procedures, reviewed the assigned aircraft Form 781 (Aircraft Maintenance Form) for discrepancies, and released the crew to perform their preflight checks.

The navigator connected the oxygen mask to the oxygen system receptacle and performed the oxygen checkout. Then he checked the periscopic sextant for timer accuracy. A parachute harness was worn but the parachute itself (a chest pack) was on board the aircraft in the rear. The pins holding the chutes folded were inspected during Preflight by opening an access panel to insure they were not bent or binding. In the event of a bailout, the parachute was to be clipped on the front of the harness just before bail out. If the navigator didn't get a "Time Hack" from the radio in the squadron, he requested the copilot tune station WWV from the National Bureau of Standards, which provided a precise time hack every five minutes. The frequencies were 2.5, 5, 10, 15, 20 and 25 MHz. Even better, the Canadian station, CHU, could be tuned, if it was in reception range, and this station provided a time hack every single minute.

Departure
An on-time takeoff was critical and was scored. A five-minute early takeoff was allowed, in case a maintenance squawk was discovered at the last minute needing to be fixed.

The reason for the criticality of takeoff time was the rendezvous time for refueling was critical; the bomber was counting on the tanker to be at the ARCP on time.

The navigator kept track of the time and announced on interphone, "Five Minutes to Start Engines," and counted down so engines were started in a precise manner. Also, the countdown was made for takeoff time. Normal training missions were not conducted at the aforementioned 175,000 lbs. gross weight, due to the marginal aircraft performance issues. Nevertheless, even at moderate aircraft weights, the KC-97 required about 45 minutes from takeoff to level off at 15,000 ft. altitude.

Early on, the standard procedure was to take off and climb South direct to Martha's Vineyard Visual Omni Range (VOR) radio navigation facility, turn east to Nantucket VORTAC (A combined radio navigation facility, combining a VOR and a TACAN, or distance measuring facility, at the same location), still climbing and turn north to a position 30 miles east of Boston VORTAC (over water). Then the aircraft flew west direct to Boston VORTAC. Just as we turned over Boston VORTAC we would be at 15,000 ft. altitude. This usually took about 45 minutes. Later on, someone figured out we were burning too much fuel to fly this "Boston Number One" departure, and we took off and climbed directly on course to the refueling area.

Takeoff Acceleration Time Check
An aircraft acceleration check was used for takeoff. The flight engineer extracted from the Aircraft Flight Performance Manual the not-to-exceed-time required to accelerate the aircraft from 60 to 100 knots indicated airspeed (IAS) on the runway, and announced this to the crew. As the aircraft approached 60 knots IAS, the copilot announced on interphone "60 Knots," and when 60 knots was actually attained, he announced "Now." The navigator started his stopwatch and as the time approached, announced "Time" and at the actual briefed time, "Hack." Both pilot and copilot verbally compared airspeed indicators (both pilot and copilot had their own airspeed meter.) As long as the aircraft accelerated within the briefed time to 100 knots, the takeoff was continued. Failure to make the time resulted in a takeoff abort. On hot days, when the air was less dense, and the aircraft was heavy, an acceleration check could be as long as 17 seconds, and it was only from 60 to 100 knots IAS!

Engineer, Set Refueling Power

Aircraft Climb to Altitude

The flight engineer had to perform a balancing act between opening the engine cowl flaps to facilitate engine cylinder head cooling and the accompanying increase in aircraft drag with cowl flaps partially open.

The engines were worked hard. Once at level off altitude, the aircraft cruised at a True Airspeed (TAS) of 220 knots. Apparently, there had been problems with the turbo supercharger bearings, because a Time Controlled Technical Order (TCTO) was accomplished, retrofitting the aircraft with turbo supercharger bearing temperature indicators at the flight engineer's station. Failure of a turbo supercharger could spew hot shrapnel all around, and this component was located near the wing fuel tanks; not a good situation. As an aside, this is as good a place as any to state the R-4360 engine failed in flight routinely, because of sustained hard usage.

Rendezvous Procedures

The tanker was navigated to the Orbit, with a planned orbit time of 15 minutes built-in to the mission, and established a left hand orbit pattern, consisting of a four minute leg, a ¼ standard rate turn to the left (four minute turn), a four minute leg and another ¼ standard rate turn to the left. At the time the tanker established itself in the orbit, the bomber was supposed to have reached the Initial Point (IP), a known fixed navigation point, inbound to the refueling track. Once in orbit, the tanker pilot attempted to make radio contact with the bomber on the briefed UHF refueling radio frequency 15 minutes prior to the Air Refueling Control Time (ARCT). Once radio contact was made, the pilot confirmed the fuel offload and weather in the refueling area, and then turned over the UHF radio to the navigator to conduct the rendezvous.

The primary rendezvous equipment was the APN-69 radar transponder. The unit was a receiver-transmitter located against the ceiling of the crew compartment, above and aft of the navigator. When the radar on the bomber interrogated the APN-69 using the Identification Friend or Foe (IFF/SIF) Mode 2, the tanker transceiver sent back an IFF/SIF coded signal (set by the navigator using his control box) readable on the bomber radar.

(Mode 2 was for military aircraft only; Mode 1 and 3 were for commercial air traffic control.) This gave the bomber radar navigator a radar range and relative bearing to the tanker. The secondary rendezvous equipment was the APN12/76. The APN-12 was in the tanker and the APN-76 was in the bomber. In this case, the tanker sent out a signal and the bomber responded, with readout at the tanker navigator panel. Unfortunately, the radar navigator on the B-47 sat on an ejection seat, and had to unstrap from the seat belt to turn around and turn on the APN-76. Obviously, he did not want to risk snagging the ejection seat lever with his parachute harness.

In the vast majority of cases, there was no signal on the APN12 in the tanker to read during the rendezvous. A backup aid to rendezvous was the UHF Radio Direction Finder. At the request of the bomber pilot, the pilot of the tanker broadcasted for 15 seconds on UHF refueling frequency, and the Direction Finder (DF) capability of the bomber radio pointed toward the tanker. No range was given in this instance, however.

Once communications were turned-over to the tanker navigator, the bomber pilot requested the APN-69 be turned to "Standby" mode, followed by "Transmit" mode. The signal on the bomber radar disappeared and reappeared during this process, allowing the bomber radar navigator to positively identify the rendezvous signal from the correct tanker aircraft. The range of the bomber from the tanker was transmitted by UHF radio at intervals of every 20 miles down to 70-mile range, every 10 miles down to 10-mile range, and not more than two miles until the one-mile range. The tanker maintained position in the orbit until the 30-mile range call from the bomber; at that point the tanker navigator directed the pilot to depart the orbit on his computed compass heading.

Radar Beacon Control Panel at Navigator Station (KC-97L)
Photo by Author

The bomber descended from 30,000 ft. and slowed down during this process, inbound from his IP. When the range of 12 miles was transmitted from the bomber, the tanker pilot announced over interphone, "Engineer, Set Refueling Power," the flight engineer set Maximum Except Takeoff (METO) power, which was 2550 rpm and maximum inlet manifold pressure (fed from the turbo superchargers). The tanker airspeed increased to 250 KTAS. The rendezvous was complete at the ARCP with the bomber in the "Observation Position."

Engineer, Set Refueling Power

As the B-47 took on fuel, and its weight increased, the receiver aircraft needed more airspeed to keep from stalling. Typically, the B-47 pilot requested an additional five knots airspeed, and the pilot of the KC-97 pushed the yoke to nose over and pick up a 500 fpm (foot per minute) descent to maintain the B-47 on the end of the flying boom without stalling. This was referred to as a "Toboggan" maneuver. End Air Refueling (EAR) frequently occurred as low as 12,000 ft., and the B-47 had to burn extra fuel to then climb back up to its 30,000 ft. altitude.

After completion of the refueling, the tanker made a 45 degree turn to the right to clear the track, and the bomber climbed straight ahead to regain cruise altitude. The tanker navigator gave the bomber an offload report. This consisted of the weight of fuel transferred (provided by the flight engineer) and an aircraft position (fix) for the bomber radar navigator to restart his navigation (The bomber radar navigator put his radar in standby mode during refueling to minimize the radar energy being transmitted, for safety reasons.)

If, for any reason, any crewmember on either the tanker or the bomber recognized an unsafe condition, he (in those days there was no "she") broadcasted on UHF radio refueling frequency, "Breakaway, Breakaway, Breakaway." The tanker pilot increased engine power, if possible, to accelerate forward, the bomber pilot dropped back and below the tanker to assess the situation.

Engineer, Set Refueling Power

KC-97G
Ref: en.wikipedia.org

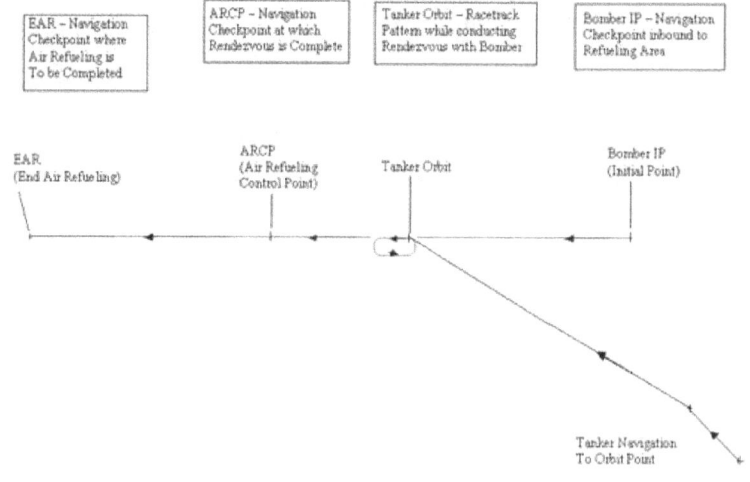

Rendezvous Procedures
Art by Author

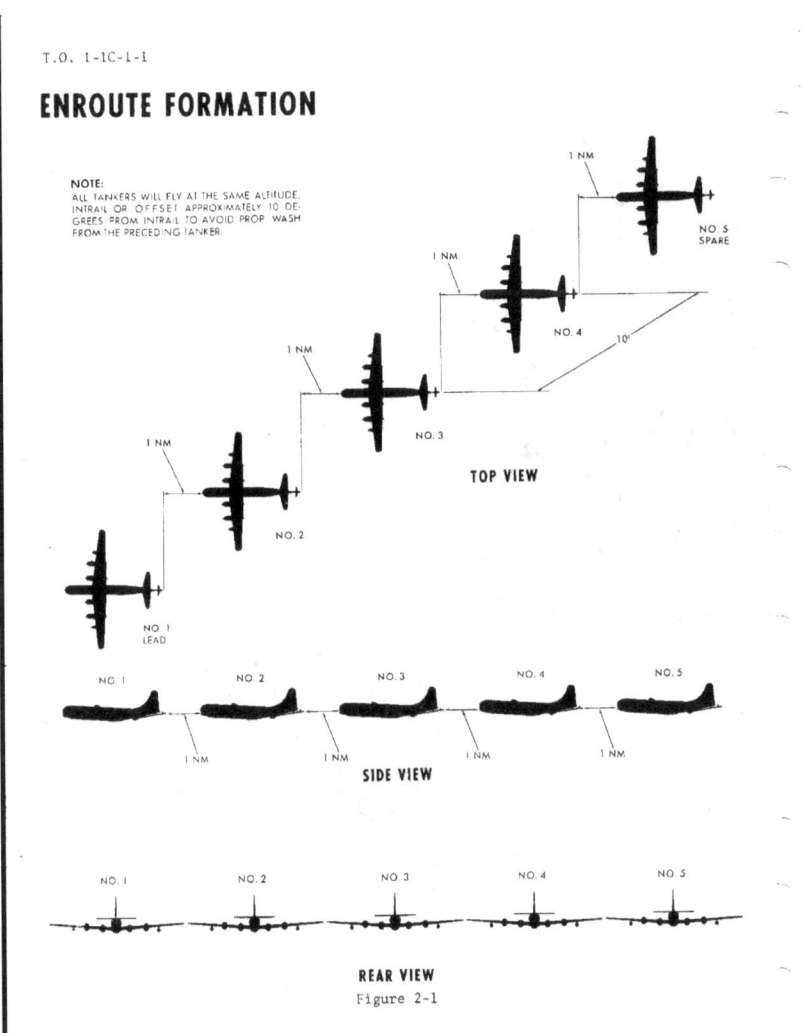

KC-97 Enroute Cell
Ref; KC-97L Flight Manual (T.O. 1C-97(K)L-1)

Engineer, Set Refueling Power

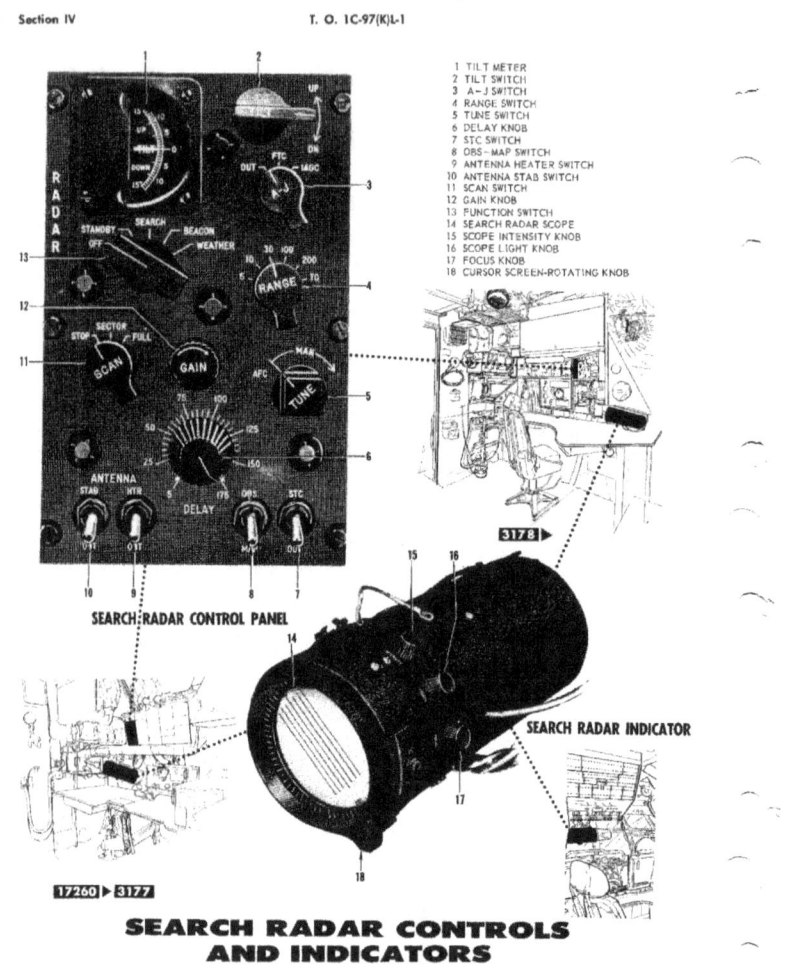

Section IV   T. O. 1C-97(K)L-1

1 TILT METER
2 TILT SWITCH
3 A–J SWITCH
4 RANGE SWITCH
5 TUNE SWITCH
6 DELAY KNOB
7 STC SWITCH
8 OBS–MAP SWITCH
9 ANTENNA HEATER SWITCH
10 ANTENNA STAB SWITCH
11 SCAN SWITCH
12 GAIN KNOB
13 FUNCTION SWITCH
14 SEARCH RADAR SCOPE
15 SCOPE INTENSITY KNOB
16 SCOPE LIGHT KNOB
17 FOCUS KNOB
18 CURSOR SCREEN-ROTATING KNOB

SEARCH RADAR CONTROL PANEL

SEARCH RADAR INDICATOR

SEARCH RADAR CONTROLS AND INDICATORS

Ref: KC-97 Flight Manual (T.O. 1C-97(K)L-1)

KC-97G Navigator Station

Close-up of Navigator's
Electronic Equipment

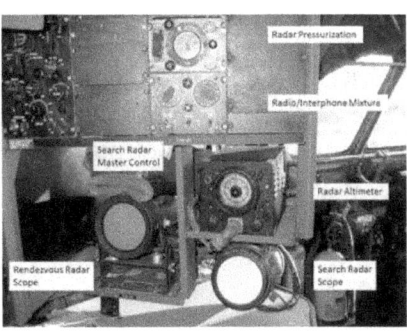

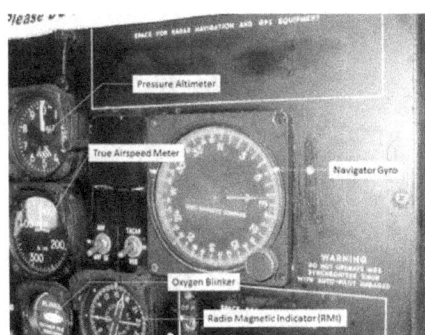

Close-up of Navigator's
Basic Navigation Instruments

Note: N1 Compass was
Canabalized

KC-97

Photos by Author

Tail Wind of 100 knots

I flew a night refueling mission, in winter, on a particularly cold night. After the refueling, I performed a night celestial navigation leg. Most of the navigation leg was due east back to Otis AFB. We picked up a slight chop (turbulence), but it was a clear night. We were at 15,000 feet altitude, cruising at 220 knots true airspeed. I made my first three-star fix, and when I plotted our position, it was about 50 miles ahead of where I projected it to be by ded reckoning. I glanced over the pilots shoulder, got a quick TACAN fix from his instrument, and, sure enough, my celestial information was accurate. We had picked up a tail wind from the jet stream, which had dipped down to our altitude in winter, and we had a tail wind of 100 knots. This was the first and last time I experienced a boost of that magnitude from the jet stream.

Single Engine Qualified Copilot *"Wanabe"*

One day, while flying enroute to the refueling area, my copilot was once again moaning about not being a fighter pilot, and the AC had enough, and told the flight engineer, "Sergeant, the copilot wants to be a single engine pilot. Give him responsibility for number two engine throttle; let him do whatever he wants with number two throttle." That took the wind out of his sails.

Thirty-six years later I tracked down, and had a reunion with my copilot; we both were retired, and he lived in Florida. He had left the Air Force after being in SAC, flew single engine fighters with the Illinois Air National Guard for a while, and eventually became a pilot (captain) with American Airlines. Unfortunately, he is gone now.

Keeping the Traffic Down in the Cockpit

During reflex deployments, we usually had passengers on board who were from the maintenance organizations. On longer flights, when boredom set in, the passengers streamed up to the cockpit to see where we were, and inquire how much longer the flight would take to destination. I didn't mind this traffic as long as it didn't interfere with my work. I borrowed an idea from another navigator in my squadron. I made an extra chart depicting the course to be flown, and taped it to one of the air refueling tanks in the main cabin. Whenever I obtained a fix of our position, I walked back and plotted this position with its time on the spare chart, so the passengers could keep up with the flight. This worked out well, and slowed the traffic into the cockpit to a trickle.

Close Call with a Tanker (Ship) in the Cape Cod Canal

I liked to go fishing, occasionally, but I did not own a boat. One of our pilots, Bruce Pike, owned a small rowboat with an outboard motor, and he liked to fish. We got together one evening, to go fishing in the Cape Cod Canal. It was dark. We sat in the boat, in the mouth of the canal, and as we caught small sand sharks, he used his hammer to knock the shark senseless. Then he went into the shark's mouth to retrieve his hook.

We also caught bluefish, and did they put up a fight. We hoped for stripers but did not have any luck. During a lull in the activity, as we sat in the boat, I had the weird feeling somebody or something was behind me. I turned around, and there was an oil tanker virtually on top of us. I nearly jumped out of the boat. The tanker passed silently just to the right of us, with no sign of having seen us. As the stern passed by, I saw the propeller turning slowly, making very little noise, and just allowing the ship to make headway going into the canal. It was a close call. The ship could have cut us in two, and never would have seen us. We took the bluefish home and cut them up for steaks. They tasted good although somewhat oily.

Abbreviated Survival School

All SAC aircrews were required to go to the USAF Survival School at Stead AFB, Nevada. That is, all aircrews except KC-97 aircrews. The reason for the KC-97 tanker aircrew exception is unknown; perhaps it was because the school was limited in the number of students it could take. When crews were reassigned to the KC-135 tanker, they were required to go to Stead AFB. My crew was sent to a local survival school set up by the SAC unit at Dow AFB, Bangor, Maine. The course lasted for four days, I recall.

We were flown up by one of our aircraft, and, when we arrived, we were told not to fill our water canteens, as there was a spring at the school site. When we arrived at the training site, in rural Maine, there was, indeed, a spring. There was a rusty steel spring suspended from a tree branch. Ha, Ha. I don't remember much about the school, except being grubby and hungry. It was tough shaving many days of beard off when I arrived back home.

The 19th ARS was Disbanded

The Air Force needed an air refueling tanker to fly at higher altitudes and at higher speed to keep up with the jet-powered bomber aircraft. A proposal was made to retrofit the KC-97 with turboprop engines, and a prototype was built. My research indicates the power plant was the Pratt & Whitney T-34 with the Curtiss Electric Propellers. It apparently was a test bed for the power plant slated for the C-133 airlifter.

Also, the option was there to add the J47 jet pods under the wings, as the Air National Guard was doing. However, the jet-powered KC-135 was in production, and the Air Force knew if an improved version of the KC-97 was available, it would not get all of its jet-powered tankers. The decision was made to retire the KC-97 fleet when sufficient KC-135 aircraft were available. Just as my Air Force commitment was up, and I was about to be released from active duty, the alert commitment for the 19th ARS was relieved. The squadron was going to be inactivated, and the unit's KC-97 aircraft were going to the bone yard. I already had my active duty separation orders and was alerted my crew had one last flight, to deliver one of our tankers to Davis Monthan AFB, the Air Force *bone yard*. I was not interested in going, but another navigator volunteered to go, so I was through flying, as far as active duty was concerned. I separated from active duty on November 15, 1965, as a captain.

CHAPTER 7; FIRST ENGINEERING JOB AT HAMILTON STANDARD

As I neared my separation date from active duty in the Air Force, I sent out resumes to several companies and government agencies. I wanted a position as an engineer in an aircraft related field. The Naval Air Engineering Center in Philadelphia invited me to interview. It went well, and I subsequently received a job offer as a Combustion Engineer at a Civil Service Rating of GS5, which paid $5990 per year. This was before the pay in civil service was raised to compete with wages paid to engineers in private industry. My job offer was appropriate for new college graduates with no industry experience. There was no consideration for my flight experience on active duty. The job would have been interesting, but, since I was making approximately $10,000 per year, including flight pay, as a captain, and had a wife and two children to support, I politely declined the offer.

I wanted to work for Pratt & Whitney Aircraft Engines (P&W) in the Hartford area. The response from P&W was to notify them when I was released from active duty, and they would determine if there was an opportunity for me at this point. I was put-off by P&W, as I needed to secure a commitment for a job before I separated from active duty, not after!

I struck pay dirt with my job application to Hamilton Standard, in Windsor Locks, Connecticut. They hired me as a Propeller Installation Engineer. My assignment was the technical interface between the engineering technical staff and customers. I routinely communicated with the Federal Aviation Agency, Lockheed Georgia and Lockheed Burbank, among other miscellaneous customers. I was assigned the 54H60 propeller system, which was the last and most sophisticated hydromatic design that started out in the 1930 time era. This propeller system was installed on the Alison T-56 turboprop-powered C-130 Hercules and the Navy P3 Orion. Both of these aircraft are still in service, as of 2018.

Background in Propellers

Propeller blades have a natural tendency to rotate to flat pitch, due to the centrifugal forces on the propeller blades. The design of the pitch control system cannot allow this, as flat pitch represents an enormous amount of drag on the aircraft, possibly resulting in a structural failure of the aircraft tail. Also, flat pitch results in an overspeed condition, which can structurally compromise the integrity of the blades. As a child, remember when you stuck a windmill fan out the window of a traveling car; the drag was quite high and the fan rotated at a high speed. This fan represented a simulated propeller at flat pitch. Accordingly, the propeller design has safety features to prevent the natural tendency to go to flat pitch.

One method, used on the 54H60 system, was to use a pitch lock mechanism, sensitive to either the loss of hydraulic pressure or propeller RPM in an overspeed condition. Another commonly used method was to bolt counterweights to the inboard section of the blade, such the blades would rotate to a very high pitch angle, near feather, in the event of a loss of hydraulic oil pressure.

Under normal operating conditions, the propeller blade angle is controlled by a governor set to maintain a selected propeller speed (PRPM). Once set, if the speed slows down, the blade pitch is reduced, allowing the propeller speed to increase to the set point. The opposite happens when the speed increases; the blade pitch is increased, increasing the air load, thereby slowing down the speed.

The propeller operates in a vibration environment, due to the air inflow angle, such as during climb. Each revolution of the propeller goes through a cycle of higher aerodynamic load to lower aerodynamic load. This is referred to as 1P vibration. Every propeller design must undergo a propeller vibration survey to measure the vibratory loading. In the operating range (speed) the vibration stresses in the blades must be below the range of values facilitating infinite life of the blade material. Frequently, the stresses exceed the allowable, as in crosswind conditions on the ground. In this case, the propeller is restricted from operating under this condition.

Propeller blade aerodynamic design is tailored to the performance requirements of the aircraft application. For example, on the 54H60 propeller, the blade design for the C-130 aircraft had a high camber, in order to maximize the takeoff and initial climb performance, and sacrifice, somewhat, cruise performance. However, on the 54H60 propeller for the Navy P3 Orion, the blade design had a low camber, sacrificing takeoff performance for optimum long-range cruise performance.

The propeller must absorb the power produced by the engine and convert this power to thrust. Isaac Newton's Second Law of Motion is F=MA, where F is the force or thrust, M is the mass being moved, and A is the acceleration given to that mass.

$$F = MA$$

In the case of a propeller, thrust is developed by taking a fairly large mass of air going through the propeller disc, and giving this mass of air a small amount of acceleration, or change in velocity (Acceleration is a change in velocity per unit of time.) This relationship applies in both cases, whether the engine is a piston engine or a turbine engine.

Propellers are rated by the amount of shaft horsepower they can absorb, called SHP. This is the amount of horsepower applied at the engine propeller shaft (on which the propeller is mounted). An engine is rated at a SHP value at a particular engine RPM, or revolutions per minute. The relationship between SHP and the torque developed at the engine shaft is as follows:

$$SHP = QN/5252$$

The value, Q, is torque or twisting force, N is the RPM and 5252 is a constant used to make sure the number comes out in foot-pounds (torque).

In the case of a turboprop engine, most of the energy developed in the engine is absorbed by the power turbine and fed back through the reduction gearbox to the propeller. There is always a small amount of residual thrust left over, exiting the engine. This small amount of thrust is converted into an equivalent horsepower, and the engine is rated by a value called Equivalent Shaft Horsepower, or ESHP. The propeller doesn't feel this extra SHP, but the aircraft does.

CC-109 Cosmo Retrofit with 54H60 Propellers

One of my early work assignments was to provide engineering support to the Canadian Forces and Pacific Airmotive in Burbank, California for the retrofit of the power plant (engine and propeller) in the Canadian Forces CC-109 twin engine Very Important Person (VIP) aircraft.

The aircraft had Napier Eland 504A power plants, and the Canadian Forces wanted the C-130 power plant, which was already in their inventory (they had C-130 aircraft) for ease of logistics. Allison had a Supplemental Type Certificate (STC) for the commercial twin engine Convair 580, which was very similar to the CC-109. Pacific Airmotive in Burbank, California was the prime contractor to perform the aircraft modification. However, the Convair 580 (Similar to the CC-109) not only had the commercial version of the Allison T-56 turboprop engine, but also had the Aeroproducts propeller (Aeroproducts was a propeller manufacturer under the General Motors umbrella, as was the Allison engine company.)

Pacific Airmotive was not happy with the Hamilton Standard selection as the propeller, nor were the Allison engine technicians on-site. I took a lot of flak from the Allison tech reps while I supported the engineering effort and covered the first engine run up and flight. It was a thrill to watch the aircraft initial takeoff, as it made a maximum performance climb to reach as much altitude as soon as possible, in case there was a problem with the propellers (there was not.) The flight test pilot made the wry comment after the first flight, "Them props were quick." He was referring to the propeller rate of pitch change going into feather, when the engine was intentionally shutdown.

Hamilton Standard 54H60-77 Propeller
Installed on US Navy P3 Orion
Ref: wikipedia

After going into service, the retrofitted aircraft experienced heavy smoke coming from the propeller synchrophaser, which was a piece of electronics supplied by Hamilton Standard, mounted in a radio rack with other electronics. As a matter of interest, the synchrophaser not only matched the RPM of all the propellers, but set up the optimum phase angle between each propeller number one blade, for reduced propeller noise and vibration.

It turns out the Canadian Forces applied a faceplate over the synchrophaser, making the support system into an oven, which restricted the natural cooling airflow over the unit, and overheated it. The removal of the faceplate solved the problem.

Tear on Photograph

CC-109 Prior to Modification

CC109 with Napier Eland Powerplants

Prototype CC-109 with Hamilton Standard Propellers Installed, awaiting a gear retraction test. Photo by Author

Engineer, Set Refueling Power

Author on entry stairway of prototype CC-109 at Pacific Airmotive. Photo by Author

Salary Adjustment

By the second summer at Hamilton Standard, I was hurting financially. I had taken this job at a loss, going from $10,000/year as a captain on active duty in the Air Force, to $8990 as an engineer. I approached my boss in a humble manner, and explained my situation. He had to wait for his supervisor to come back from vacation, and then they both huddled. They decided I had hired-in at too low a salary, and made an immediate adjustment. I felt I was treated fairly, and was happy.

C-130A Propeller Retrofit

I was assigned to support the retrofit of the C-130A with the later model propeller, the Hamilton Standard 54H60. As background, the C-130A was initially manufactured with the Allison T56 engine and the Aeroproducts propeller in the early 1950s.

This first model of the C-130 did not have a low-speed ground idle setting on the engine fuel control, so the engine screamed constantly (this was a constant speed engine), even when taxiing on the ground. The Aeroproducts propeller was a three-bladed design, using hollow steel blades of 15 feet diameter. The gear ratio in the propeller gearbox reduced the propeller RPM to 1107 RPM. The noise level was sufficiently loud, ear damage was a consideration for the crew. Subsequently, a design change was made on the C-130A, to lower the engine RPM, such as the propeller ran at 1020 PRPM.

On the next version of the C-130, the C-130B and later model E and H, the aircraft used the four-bladed Hamilton Standard 54H60 propeller. This design had a reduced diameter, 13.5 feet, and ran at the lower propeller speed of 1020 RPM. Both the reduced diameter and lower propeller speed helped lower the noise (reduced propeller blade tip speed). Also, the later model aircraft had a low-speed ground idle setting on the engine coordinator, dropping the engine RPM to approximately 10,000 on the ground from 13,820 in flight. These changes reduced the aircraft noise level for the crew and passengers, but the aircraft was still considered noisy.

The C-130A was equipped with Aeroproducts three-bladed propellers, of 15 feet diameter. The author was responsible for the initial two prototype conversions to the Hamilton Standard 54H60-91 propeller system. Aeroproducts had already gone out of business, and the Air Force was concerned there would not be sufficient propeller assets to support the C-130A fleet worldwide, including Vietnam.

Lockheed C-130A with Aeroproducts Three-Bladed Propellers
Ref; nationalmuseum.af.mil

Engineer, Set Refueling Power

The C-130 engine, the Allison T-56, could be considered a second-generation turboprop engine. It was what was known as a "solid shaft engine." The entire engine was connected mechanically through shafting. The engine compressor was shafted to the high-pressure turbine, which was connected to the low-pressure turbine, which was connected to the propeller through a reduction gearbox. The engine ran at one speed, 13,820 RPM, with the exception of ground idle and the *beta* mode.

The *beta* mode was a schedule of propeller blade angle and engine speed as set by the throttle while taxiing on the ground. Once flight idle and higher power was selected with the throttle, the speed was stepped up to 13,820 RPM/1020 propeller RPM. Most modern engines have gotten away from the single speed engine concept, by incorporating a low-pressure turbine as a free turbine. This means the low-pressure turbine is not connected to the high-pressure turbine by a shaft; rather the two turbines are gas coupled only.

The Vietnam War was raging in the late 1960s, and the C-130 aircraft was the prevalent aircraft type used for forward unprepared airstrip resupply. Using assault-landing procedures, the aircraft typically approached at a high speed and dropped on the runway fairly hard with full propeller reverse thrust and hard braking. This kicked up dust, stones and whatever, and this mass of foreign objects eroded the propeller blades quickly. At this time, Aeroproducts had just gone out of business, and the Air Force was concerned it would run out of propeller blade assets for the C-130A due to the hard erosion usage in assault landings. The thought was to retrofit the C-130A with the later Hamilton Standard 54H60 propeller system, which was still in production.

A team was put together, consisting of aircraft electricians and sheet metal workers, from the Pratt & Whitney Experimental Hangar in East Hartford. I was the Project Engineer. I made the airline reservations for the team to travel to Lockbourne AFB in Columbus, Ohio, where the retrofit was to take place. There was an operational C-130A squadron located at Lockbourne and arrangements were made to retrofit one of their aircraft. P&W, in those days, had an agreement with the airlines; the airlines paid first class prices for the commercial P&W engines, and P&W personnel traveled first class when flying with the airlines. So, when I arrived at the airport to travel to Lockbourne, I found out I, as the project engineer, sat in coach, and the blue-collar workers sat in first class! P&W had changed the tickets from coach to first class, without my knowledge.

The field retrofit took place in the summer of 1967, and took approximately three months. In addition to the mechanical hanging of the propellers on the T-56 engine shaft, the retrofit consisted of completely rewiring the aircraft propeller circuits to incorporate the propeller feathering circuitry through the engine shutdown controls, the deice/anti-ice circuitry, and the synchrophaser wiring. Engineering drawings were prepared to define the retrofit and were presented to the Air Force Warner Robins Air Material Center, so other aircraft in the C-130A fleet could be retrofitted on a production basis by another contractor.

The author helping to run propeller wiring inside the prototype C-130A rtrofitted with the 54H60 propeller.
Photo by Author

"Pseudo" Problems after Retrofit

One problem noted by the flight test pilot, while testing the first retrofitted C-130A, was the aircraft taxied too fast with the 54H60 propellers installed. Of course, no flight test was accomplished before the retrofit for comparison. This didn't make any sense to me, and I requested the aircraft throttle rods/bell cranks be rigged properly, from the cockpit throttles all the way out to the wings and nacelles to the engines. A civil service technician was sent from Warner Robins AFB to accomplish this rigging. After this was accomplished, a repeat of the taxi test yielded the same results; too much speed resulting in using too much braking to control the aircraft on the ground.

I solved this problem by proving the engine control rods and bell cranks were completely worn out, and had too much hysteresis. In other words, when the pilot pushed the throttle slightly to get the aircraft moving, and then pulled the throttle back to the original throttle angle, the engine control didn't return completely to the original angle. So the engine control "thought" the pilot was asking for more engine speed than was the case. I really sweated this problem out, and was pleased to be able to solve it.

Then another problem surfaced. While the Air Force flight mechanic ran the engines on the ground, the engine RPM wandered outside of limits (the engine ran at a constant speed, 13,820 RPM, with a small tolerance.)

He immediately pointed the finger to those *new* propellers. About this time, I was introduced to the civil service Air Force tech rep. I explained to him why I thought the problem was in the particular T-56 engine's fuel control. (The primary control of engine RPM was the propeller governor, but the secondary control of RPM was the engine fuel control governor.) I suggested he pitch lock the propeller, by activating the Fuel Governor Topping Switch in the cockpit, thereby intentionally resetting the propeller governor to 103% speed, and engaging the propeller pitch lock. This was a valid maintenance procedure to lock up the propeller pitch, thereby preventing any pitch change from influencing the engine RPM. Sure enough, the engine RPM continued to wander, thereby confirming the problem was in the engine fuel control, not the propeller. It was more difficult to change an engine fuel control than the propeller control, but once the engine fuel control was changed, the problem was solved. I am glad I was able to solve this one, also.

Second Retrofit to verify Aircraft Performance

Once the retrofit was completed, an Air Force crew started a limited flight test to backfill the aircraft performance data in the Flight Manual of the C-130A aircraft with the later model 54H60 propellers. A problem quickly arose, as the Air Force simply wanted to add this performance data as a direct comparison to the aircraft with the Aeroproducts propeller. This would have used a sample of one to compare with original flight test data based on average data for the original configuration. The solution was to conduct an aircraft performance flight test on an unaltered C-130A, accomplish an additional retrofit on this aircraft, and repeat the aircraft performance flight test on this specific C-130A with the 54H60 propellers.

The Hamilton Standard Overhaul and Repair facility was tasked to do the second retrofit, as described above. This was scheduled at Edwards AFB, California. I was asked to go along as a consultant, as I had accomplished the first retrofit. Again, this took about three months, but the results were satisfactory. The Air Force then made the back-to-back aircraft performance evaluation. Hamilton Standard made a ton of money manufacturing 54H60 propellers to retrofit the C-130A fleet.

Competition for the Air Force AX Attack Aircraft

In 1969, the Air Force issued a Request for Quote (RFQ) for a new attack aircraft. Hamilton Standard worked with the competing airframe manufacturers to propose a propeller, as the company was convinced the best power plant for this aircraft was a turboprop. In the end, the Air Force chose a turbofan-powered design, with the GE TF34 engine. The aircraft was designated as the Fairchild A-10. With the loss of this contract, Hamilton Standard started layoffs, and

Engineer, Set Refueling Power

I was convinced it was not in my best interest to remain in the propeller business.

My Attempt to Transfer to Pratt and Whitney Aircraft Division

A peer and friend in engineering was a former Air Force pilot, and he pushed me to join the Air Force Reserve at Westover Air Force Base, in their C-124 aircraft (See next chapter). Somehow, my friend managed to get a coveted transfer to Pratt and Whitney, as a flight operations engineer. I was jealous, as this job seemed right along with my career interests. I was quite vocal about not believing propellers were the wave of the future, and, when I approached my supervision about looking for a job at P&W, I was told in no uncertain terms "No, not until I changed my belief about propellers." It was obvious I was going to have to leave the company. In this time frame, there was no job mobility between United Aircraft divisions, unless the company initiated the process.

My Interview with Lycoming

Lycoming, a jet engine manufacturer located in Stratford, Connecticut, ran an ad in the Hartford Courant, for engineers. I took an interview, and it went like this, "Do you want to go to Vietnam? We are hiring service engineers for the Huey Helicopters, and need in-country engineers to service the engines." They were hiring anybody with an engineering degree and a body temperature of 98.6F. Heck, if I wanted to go to Vietnam, I would have stayed in the Air Force. I declined interest in the job.

Aero Spacelines Super Guppy with 54H60 Propellers

Hamilton Standard received a contract from Aero Spacelines to support the conversion of one of their commercial B377 aircraft from the Pratt & Whitney R4360 piston engine to the commercial T56/54H60 turboprop powerplant. Again, I was called upon. I accomplished all the engineering work, and made a visit to their Santa Barbara, California facility. The aircraft had a bulbous upper fuselage for carrying outsized cargo, and a nose hinged to allow frontal loading. About this time I was in the throes of seriously looking for another position. Aero Spacelines had a job posting in the Hartford Courant for a Powerplant Engineer. I submitted my resume through the normal process, by mail. This was a foolish move, as I had personally met the new vice president of the company on my business trip to Santa Barbara. The job application process took some time, and I had already accepted a job with GE.

One evening, while I packed up my household items in preparation for my move to GE in Cincinnati, Ohio, I received a telephone call from the vice president of Aero Spacelines, indicating he just saw my application for his company's open job.

Aero Spacelines Super Guppy
Ref: http://en.wikipedia.org/wiki/File:SuperGuppy-F-BPPA.jpg

He tried to sell me on the job, even offering to fly my wife and me to Santa Barbara to observe the good life in that area. I had to politely decline, as I had already accepted the job with GE. The smart thing would have been to contact this man when the job appeared in the Hartford Courant, but I had not learned this lesson in life, yet. I always refer to this incident as my near miss to move to Santa Barbara, California.

My effort to support the application of the 54H60 propeller on the

Engineer, Set Refueling Power

Aero Spacelines Super Guppy was a fun job, but I left Hamilton Standard before the aircraft entered flight test. Later, when working for GE in Cincinnati, Ohio, I saw a photograph in the Los Angeles Times May 13, 1970 issue, of the burned out carcass, after the prototype aircraft crashed and burned on takeoff, killing the whole crew. The cause of the accident had nothing to do with the propellers. As an aside, another factor in honoring my commitment to join GE was this company was solid financially, whereas Aero Spacelines was shaky at the time.

Aerospace engineers were valued assets in the timeframe (1965 through 1969) I worked at Hamilton Standard. The company provided annual salary reviews (with raises in the engineers' early years), and an annual cost of living adjustment.

With 20/20 hindsight, my job at Hamilton Standard exposed me to upper management in a way I never accomplished at GE. I probably should not have left Hamilton Standard, but I would not have known about the exposure issue if I stayed. As I was leaving, Hamilton Standard was having layoffs, and raises were out of the question. My boss told me I made his job easier by leaving, as there was to be only one raise in engineering that year, and I was scheduled to receive it!

CHAPTER 8: AIR FORCE RESERVE AND C-124C EXPERIENCE

Recruited for the Air Force Reserve (AFRES)

While working for Hamilton Standard, winters in Hartford, Connecticut were boring. I had nothing to do weekends, but watch football on television. Since I was not a football fan, this diversion was not exciting.

An engineer peer at HSD had been a pilot on active duty in the Air Force, and was now flying for the AFRES. He told me about the unit at Westover AFB, which was about 30 miles away. The 337th Military Airlift Squadron (337th MAS) had just retired their C-119 Flying Boxcar aircraft to Davis Monthan AFB (the Air Force "boneyard") and received eight C-124C Globemaster II long-range cargo aircraft. The authorized crew manning was increased from 1.5 crews per aircraft to 2.0. The unit was actively recruiting pilots, navigators, flight engineers and loadmasters to fill out their authorized manning.

Air Force Reserve 337th MAS turned-in their C-119 Flying Boxcars to the "boneyard."
Ref: commons.wikimedia.org

Mission

The C-124 squadrons were equipped with eight aircraft each, and were manned with two combat ready crews per aircraft. In other words, there were 32 pilots, 16 navigators, 16 flight engineers and 16 loadmasters, checked out and fully qualified to fly the aircraft. The mission of the aircraft was to fly outsized cargo in support of the active duty Air Force. Typically, the aircraft was flown to a base with an Airlift Command Post (ACP) and took on a cargo mission. This mission could be anywhere in the world. The 337th MAS supported the war in Vietnam by sending two aircraft per month to Dover AFB to enter into the Military Airlift Command system, pick up a cargo load and go. I will go into this in more detail later in this chapter.

I was reluctant to sign-up because the war in Vietnam was raging; it was 1967. I wasn't interested in a recall to active duty. An Air Force Reserve lieutenant colonel, who was in charge of recruiting, called me. He said the Air Force had a policy to strip me of my wings if I didn't join an active reserve unit within one year of leaving active duty. It turns out he was **lying**! So much for an officer and a gentleman.

This "recruiter" told me he reserved a "navigator slot" for me through the next weekend, which was the regular training weekend. He invited me to come to Westover and look over the unit, which I did.

The unit's navigator Air Reserve Technician filled-me in on the details on joining the reserve. He gave me the particulars and then we went on-board the C-124C aircraft. The first thing I noticed was the navigation equipment was the same as on the KC-97. The exception was the Loran electronics which was the older WWII technology APN9 set.

Engineer, Set Refueling Power

Air Force Reserve 337th MAS was assigned eight C-124C Globemaster II aircraft.
Ref: commons.wikipedia.org

While at navigator school, we were taught to use the APN9, but were cautioned we would never encounter the equipment on active duty, as the more modern APN70 was in use. As an aside, there were two electronic traces on the scopes of the Loran sets. The object was to match the "Master" and "Slave" traces, one on top of the other, until the final matching, when they were superimposed over each other. When the traces were on top of each other, it was necessary to remember which trace was the master and which the slave. It was opposite between the older APN9 and the more modern APN70 sets. The way to remember was to impose sex into the equation. The old set (APN9) had the Slave (Sam) on top of the Master (Mary). The newer set (APN70) had the Master (Mary) on top of the Slave (Sam). Once it was explained this way, nobody ever forgot which trace was which.

The cockpit was at the top of an approximate 15 foot long ladder from the cargo compartment. This ladder was removable to facilitate the loading and unloading of cargo through the front clamshell doors.

Cargo bay, forward looking aft. Upper deck folding floor stowed to load oversized cargo
Ref: commons.wikimedia.org

Navigator's Driftmeter

The cockpit on the C-124 aircraft was configured about 20 feet off the ground, over the clamshell doors and drive-in loading ramp. This meant the navigator's driftmeter had to be located outside the cockpit. The driftmeter had a fixed length tube through the aircraft, to enable the navigator to sight on objects flowing beneath the aircraft, and measure the instantaneous drift from the in-flight wind. The driftmeter was located in the cargo hold. In order to use the driftmeter, the navigator checked off interphone, climbed down the 15-foot ladder from the cockpit to the cargo hold, took his reading, and climbed back up. Let us say, I rarely used the driftmeter.

Engineer, Set Refueling Power

Photo by Author
Aft looking forward, with the upper floor deployed in order to accommodate 200 troops.
Note the ladder up front to the cockpit. Behind the ladder are the folding drive-on ramps.
The ladder was removable to load cargo.

Forward looking aft
Only two upper deck panels deployed. The folding curtains in the rear are around the
latrine. Photo by Author

C-124C Navigator Station
Ref: amcmuseum.org

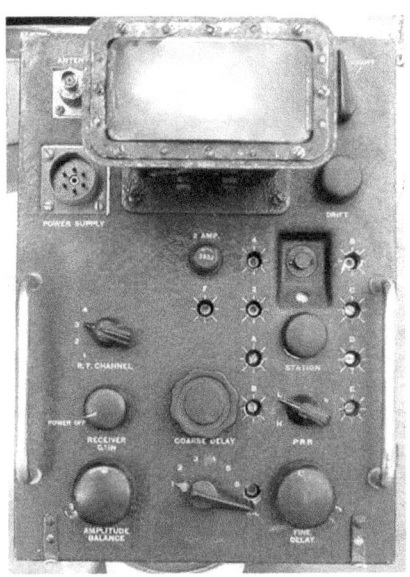

APN9 LORAN used super heterodyne "vacuum tube" technology. There was no cooling system, so one had to be careful not to leave the set on too long or it would overheat and go out of calibration. The screw driver slots provide access to recalibrate the set, as necessary.

Engineer, Set Refueling Power

Funct.	Stn.	PRR	SCOPE PATTERN	Remarks	Adj.	Crse.	Fine
			AN/APN-9 LORAN CALIBRATION				
5	0	L or H		Adjust for distinct picture	Focus		
5	0	L or H		Adjust to operator's desire	Brill		
1	0	L or H		Adjust alternately for length & position	Sl. Swp. H Cent & Ampl.		
4	0	H	1 2 3 4 5 1 2 3 4 5	5 (1000 Spaces) between (5000 Mkrs)	C		
4	0	H	1 2 3	3 (5000 MS Markers)	D		
4	0	L	1 2 3 4	4 (5000 MS Markers)	E		
4	0	H		Marker reads between 11,000 & 11,500	5th hole Right Side	C.C.W.	C.W.
4	0	H		Marker just off screen recheck previous step	2nd hole Right Side	C.W.	C.W.
4	0	L		Marker reads between 13,500 & 14,000	4th hole Right Side	C.C.W.	C.W.
4	0	L		Marker just off screen Recheck previous step	3rd hole Right Side	C.W.	C.W.
2	0	L or H		Adjust alternately for length & position	Fa. Swp. H Cent & Ampl.		
2	0	L or H		Adjust to center vertically	Vert. Center		
5	0	L or H	1000 MS MARKER 100 MS MARKER	Adjust 1000MS mkrs $\frac{1}{8}$"-$\frac{1}{4}$" above 100MS Mkrs	Marker Ampl.		
5	0	L or H	CROSS HAIR	Adjust until cross-hair barely touches 10 MS markers on upper trace	Cross Hair		
5	0	H	Bottom Trace 1 2 3 4 5 / 50 40 30 20 10	4 (10 MS Mkrs) between (50 MS Mkrs)	A		
5	0	H	Top Trace 1000 MS	9 (100 MS Mkrs) between (1000 MS Mkrs)	B		
5	0	H	Cross Hair 1 2 3 4 5 6 7 8	Count 8 spaces from cross-hair on left to first stn rate on right (BTM trace)	B		
5	1	H	Cross Hair 1 2 3 4 5 6 7	Count 7 spaces from cross-hair to STN rate marker	B		
5	2	H	1 2 3 4 5 6	Count 6	2		
5	3	H	Cross Hair 1 2 3 4 5	Count 5	2		
5	4	H	1 2 3 4	Count 4	4		
5	5	H	1 2 3	Count 3	4		
5	6	H	Cross Hair 1 2	Count 2	F 6		
5	7	H	Cross Hair 1	Count 1	F 6		
5	4	H	Move L-R switch to right, stn. rate marker should jump two spaces left		B R-L		
5	0	H	Rotate fine delay C.W. to C.C.W. and count hundreds; read no less than 700 nor more than 1500		1st hole Right Side	Set on 5000	C.W. & C.C.W.

Elaborate Calibration Procedure if the APN9 LORAN set goes out of calibration

Engineer, Set Refueling Power

After the aircraft tour, we went to the room where the navigators hung out. This unit was under the Military Airlift Command (MAC) umbrella. The mission was to support the active duty force with airlift capability. In order to fulfill this mission, there was a blackboard with a listing of destinations in white chalk. Listed were typical trips, such as the following:

Rhein-main AB, Germany September 7-13, 1967

Yakota AB, Japan, October 15-22, 1967

Lajes AB, Azores, October 24-29, 1967

The Air Reserve Technician told me these were the cargo missions the squadron was committed to fly, in support of the active Air Force.

I said, "How do I get on the trip?"

He said, "Put your name next to the date in chalk."

Since Marilyn was positive about the idea of my flying again, I signed up. I was excited. On active duty, I had limited travel opportunities. I pulled alert at Sondestrom AB, in Greenland, traveled to England and California. Now, I had the opportunity to travel worldwide, and be paid at the same time. Then, there was the carrot of retirement. It wasn't until I was almost ready to retire did I realize how important the Air Force retirement benefit was.

When I was a navigator on active duty in SAC, I was responsible for the tactical procedures to accomplish the rendezvous between the tanker and bomber, as well as general navigation. In the MAC C-124C, I was responsible for general navigation as well as plotting fuel burn vs. distance flown. I planned the mission as before, and extracted the fuel required from the fuel burn manual for the aircraft. I then plotted a "planned" line on the fuel burned chart. At each navigation fix, the flight engineer provided me with the fuel burned and I plotted that value vs. distance flown on the chart. I had to insure the actual fuel-burned line was below the planned line.

On active duty, SAC utilized the integral crew concept. I was assigned to a crew consisting of pilot, copilot, navigator, flight engineer and boom operator. We trained together, flew missions together and pulled alert together. However, in MAC, crews were put together with whoever was available at the time.

Checkout as Navigator

My first trip was to Rhein-main AB, Germany to check-out as a line navigator. I was excited about flying once again, and forgot my civilian clothes. It was about halfway to Westover AFB when I realized my civilian clothes were left at home. Rather than return home, I decided to continue without civilian clothing.

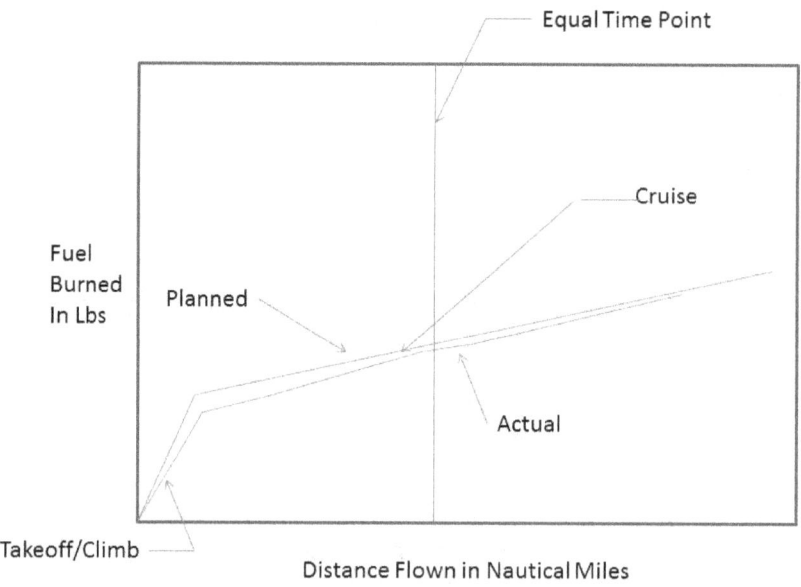

The piston powered C-124C had to island hop, with overnight crew rest in between stops. The itinerary was Westover AFB to McGuire AFB New Jersey, to Argentia, Newfoundland, to Goose Bay AB, Labrador, to Keflavik AB, Iceland, into Rhein-main West Germany. This destination was co-located at Frankfurt International Airport; the base had been a former Luftwaffe base during WWII. The return trip departed Rhein-main AB, to Lajes AB Azores, to McGuire AFB, New Jersey, and into Westover AFB. The total flying time was 45 hours.

I suffered with a toothache before we left Westover AFB. I checked with the base dental clinic, but the dentist could not find anything wrong with my teeth. I suffered all through the trip to Germany. It was my turn to perform my navigation checkout leg from Rhein-main AB to Lajes AB in The Azores. It was about 0230 hours (2:30 A.M.) local time when we were on final approach to Lajes. I called the pilot on interphone and told him about my toothache and how I couldn't stand it anymore. He called Lajes and requested the base dentist meet the aircraft. He took me to his clinic and opened it up for me. After his exam, the dentist said he could not find anything wrong, but I probably had a dead nerve. He gave me a choice; either have him pull the tooth or wait until I got home to have a root canal performed. I had him pull the tooth.

During the whole trip, the problem with my tooth affected my appetite. I had lost a few pounds by the time we landed in Lajes. After landing, the flight engineer determined we needed an engine change before proceeding. We had to wait for a part to be flown-in, so I had an extra day off. It was then I decided to buy a civilian suit in the base exchange. It fit me without alterations. Later, after arriving home and gaining the weight I had lost, the suit never fit me again.

"Brass Monkey, Brass Monkey, Brass Monkey"

When flying in West Germany, aircrews had to be cognizant of the Buffer Zone dividing West and East Germany. The Soviets liked nothing better than having the opportunity to shoot down a U.S. military aircraft for a Buffer Zone violation. If a ground based radar observed an aircraft flying toward the Buffer Zone, it broadcasted on Guard frequency, **"Brass Monkey, Brass Monkey, Brass Monkey."** Upon hearing this, all aircraft were to obtain a positive fix on their location. If it could not be done, the aircraft was to turn to a compass heading of 240 degrees, which was a safe heading to insure a "lost' aircraft would stay in a safe area.

Free Time at Rhein-main AB

After arrival, the crew chomped at the bit to go downtown to a beer hall, which was adjacent to the Haptbaunhof (Main Train Station). The rest of the crew wore civilian clothing, but I had to wear my blue uniform. I felt conspicuous. I had the only blue uniform in the beer hall. An elderly local came up to me and attempted to engage me in conversation. He spoke only German, and I could not understand him, but he kept using the word, Luftwaffe. I got the picture he was in the Luftwaffe in WWII. I was out of place. Later, I found out the U.S. military discouraged their forces from wearing their uniforms off-base, in order to minimize their presence in West Germany. Later we went to a local restaurant. I had *Veal a 'la' Gina Lolabrigida*. It consisted of a grainy piece of meat with two peach halves and two cherry halves on top. It was good and the name was appropriate.

C-124C on display at the Air Force Museum
The cowling is open on the engines.
The clam shell cargo main loading doors are also open with the drive-on ramps deployed.

Background on the C-124C Globemaster II

The aircraft was built by Douglas Aircraft in the early 1950s and initially saw service during the Korean War. The piston engines were Pratt & Whitney Aircraft R4360s. These incorporated 28 cylinders developing 3800 shaft horsepower each, and drove a Curtiss Electric three-bladed propeller of 17 feet diameter. This propeller incorporated constant speed, full feathering and reversing, and a synchronizer to match the engine speeds to minimize vibration. Cargo was loaded both from the front open clamshell doors with drive-on ramps and via a mid-ship cargo elevator.

The following is quoted from Wikipedia:

"Douglas Aircraft developed the C-124 from 1947 to 1949, from a prototype they created from a World War II–design Douglas C-74 Globemaster, and based on lessons learned during the Berlin Airlift. The aircraft was powered by four large Pratt & Whitney R-4360 piston engines producing 3,800 hp (2,800 kW) each. The C-124's design featured two large clamshell doors and a hydraulically actuated ramp in the nose as well as a cargo elevator under the aft fuselage. The C-124 was capable of carrying 68,500 lb (31,100 kg) of cargo, and the 77 ft (23 m) cargo bay featured two overhead hoists, each capable of lifting 8,000 lb (3,600 kg). As a cargo hauler, it could carry tanks, guns, trucks and other heavy equipment, while in its passenger-carrying role it could carry 200 fully equipped troops on its double decks or 127 litter patients and their attendants. It was the only aircraft of its time capable of transporting heavy equipment such as tanks and bulldozers without disassembly.

Douglas C-74
Ref: Public Domain, https://commons.wikimedia.org/w/index.php?curid=132711

The C-124 first flew on 27 November 1949, with the C-124A being delivered from May 1950. The C-124C was next, featuring more powerful engines, and an APS-42 weather radar fitted in a "thimble"-like structure on the nose. Wingtip-mounted combustion heaters were added to heat the cabin, and enable wing and tail surface deicing. The C-124As were later equipped with these improvements."

Crew Stations
The navigator and flight engineer sat behind the pilot and copilot, respectively, in the cockpit.

Life as a Reservist
I was in a Category "A" unit, which meant I was required to attend one weekend for training per month. This was referred-to as a "drill," or unit training assembly (UTA). We were compensated for one day base and flight pay for each four hour block of time. So, a weekend was paid as four days base and flight pay. We normally didn't fly during a drill weekend. Rather, we accomplished all our ground training requirements, such as annual flight physical, records

check, immunizations, water survival training in the base pool, etc.

Additional Flying Training Periods (AFTPs)

All aircrew personnel were authorized 32 annual AFTPs per year. This was how they normally accomplished their aircrew flight training. In order to qualify for an AFTP, they had to accomplish a four-hour duty day and average 2.5 hours of flight per quarter. I used the word average, as there were times the aircraft broke down before finishing the 2.5 hours. As long as we made up for the shortage over a quarter, we were paid. Our compensation was, again, one day base and flight pay.

Flight Engineer station
Photo by Author

The squadron scheduled a flight to Bermuda and back (without landing) many Friday nights, which counted as two AFTPs. We reported for flight planning before 2000 hours (8:00 P.M.), took off by 2130 hours (9:30 P.M.), landed after 0230 hours (2:30 A.M.) Saturday, and left the base by 0400 hours (4:00 A.M). The idea was to perform a minimum of a four-hour duty day on Friday and the same for Saturday, while flying a minimum of two and a half hours each day. This way, we were paid two day's base and flight pay for one flight. In 1967, this represented about $34/day, net, or $68 for the flight. This was an easy way to earn some extra money.

I made this flight a few times, most by sleeping in the cargo hold. This was in the days when navigators could log flying time whether in the navigator seat or not. Pilots had to actually be in the seat or performing duties as an instructor pilot, in order to log flying time. Later, after I left flying, the same requirements were levied on the navigator for logging flying time.

Short Active Duty Tours
We also participated in short two and three-day active duty tours. These usually ran from Friday through Sunday. We were paid the full complement of compensation; base, flight, subsistence and housing allowance. In addition we were paid a temporary duty allowance.

Annual 15 Day Active Duty Tour, or "Summer Camp."
As reservists, we were required to perform an annual 15 day active duty tour. We could sign-up for any trip the unit was scheduled for. One year, I signed up to fly to Vietnam and back. Another year, I flew two round trips to Lajes AB in The Azores. We were paid the full complement of compensation for these trips; base, flight, subsistence, housing allowance and TDY pay.

In addition to trips to Vietnam, the 337th MAS supported the active force by flying cargo all over the world. Trips to Germany, Spain, and The Azores were frequent. Also, there were three-day trips to Bermuda and Puerto Rico.

Reserve Activity Credit toward Retirement

Twenty "good" years are required for a military retirement. A "good" year consists of accumulating a minimum of 50 "points." Each duty day was one point. The UTA was a weekend "drill" but it was credited as four blocks of four hours each. So a UTA counted as four points. Similarly, an AFTP was one point. The following is a typical year:

	Annual Points
12 UTAs, or 48 points	48
32 AFTPs, or 32 points	32
15 Day Active Duty Tour, or 15 points	15
Gratis for participating	<u>15</u>
	110 points

Usually, there were at least a couple of short tours in any one year, so each tour probably was for three days, so six days are added to the total;   116 points for the year

This year represents 116/360 years toward retirement, or 0.322 years toward retirement, for calculating retirement pay.

Most aircrew members had at least five years on active duty, so they started out with 1800 points. Added to this would be 15 years at 116 points/year, and at 20 years the total would be 3540 total points.

Reservists could retire with 20 year's service, but could not start collecting their military pension until age 60. Upon reaching this age, retirement compensation was calculated by adding up the point total, dividing by 360 and multiplying by the base pay in effect at this time.

Example:
An aircrew member retired as a lieutenant colonel with over 20 year's service. The base pay in effect in 2017 is $8797.83/month.

The total number of years of service toward a reservist retirement is as follows:

Total Points of 3540; total credited years is 3540/360=9.83 years of credited service

The gross monthly pension would be 9.83/20 x $8797.83 = $4324.13

Support of Air Force Airlift Needs to Viet Nam

Our squadron was scheduled to fly two round trips to Vietnam per month, hauling cargo to support the active duty force. If a crewmember had the time available, he could sign up for more than one tour to Vietnam per year. Some crew members with seasonable employment, such as school teachers, used this means to supplement their income.

I signed-up for one of these trips in March 1969. The routing was from Westover AFB, Massachusetts, where the unit was based, to Travis AFB, California, where we picked up a cargo load heading west. From there the itinerary was as follows:

> Hickam AFB, Hawaii, crew rest and refueling; 12 hour layover for rest and relaxation

> Wake Island, overnight, crew rest and refueling

> Anderson AB, Guam, just refueling

> Mactan AB, Philippine Islands, overnight, crew rest and refueling

> Cam Ranh Bay AB, Vietnam, discharge cargo, upload cargo returning to the U.S.

The return trip made refueling and crew rest stops at the same locations. I was impressed with Hawaii. It was the most beautiful place I had ever seen, and I resolved to return as a civilian on vacation. As far as I was concerned, if God was to give the earth an enema, he would stick it in anywhere west of Hawaii.

Foolishly, I assumed we would fly all day, crew rest at night, and so forth. This should have been a piece of cake. However, after the first leg of the journey, it was fly all night, sleep all day. This was definitely not what I had envisioned.

Upon return to Westover AFB, I didn't want to see the inside of a C-124 for a while. We traveled 20,000 nautical miles round trip in 102 hours of flying time, in 10 days.

C-124C Globemaster II
I took this photo while standing up through the overhead escape hatch of our C-124C. The aircraft in the foreground was an active duty C-124C. I believe this photo was taken on the ramp of Wake Island.

Wake Island
There is a runway and some buildings, including the Drifter's Reef, which was an outdoor bar facing the water. I saw a WWII leftover "pillbox" facing the water, with a sign indicating there was unexploded ordinance in the ground, and the area was off limits.

While passing through Hickam AFB, I was impressed with the respect accorded members of the military. The copilot on our crew, a New York City detective, was able to cash a personal check in Honolulu without any problem, using his military ID card for identification. This was unheard of on the mainland. I recall going to a Mumu factory and buying a couple of Hawaiian shirts. These shirts wore like iron and lasted many years.

Filming of the Movie, "Tora, Tora, Tora."
During our stop at Hickam AFB, on the way to Vietnam, I bought something in the base exchange. As I exited the building, a flight of North American T-6 single-engine aircraft flew by, low, in formation, painted to look like Japanese Zeros. The movie "Tora, Tora, Tora" was being filmed, and these aircraft were simulating the Japanese attack aircraft. One elderly woman looked up, saw the aircraft with Japanese symbols painted on the fuselage and wings, shook her fist and shouted, "Bastards." It was odd, but I envisioned she had lived through the original attack in December 1941.

"Everyone is Trying to Kill Me"

The pilot on our crew was our unit's squadron commander. In civilian life, he was a telephone installer for the Bell Telephone Company in New York City. He was somewhat nervous all the time. Upon boarding the aircraft, while the rest of the crew ran their Preflight Checklists, he double checked everyone's work. First it was the flight engineer, as he double checked the fuel gages for the fuel on-board. Then it was my turn, as he examined my navigator's flight plan and fuel planning calculations, in detail. All the while, he muttered under his breath, "Everyone's trying to kill me. Everyone's trying to kill me." After takeoff and climb out, once the autopilot was engaged, he settled down and acted normal until the next leg of the route. Then it was the same routine all over again. I was anxious any time he was around me.

Mitsubishi A6M4 Zero
Ref: en.wikipedia.org

Note similarity of T-6 to Japanese Zero

T-6 Texan
Ref: commons.wikimedia.org

Mission Planning

My flight plan included the selection of the check points along our route, plotting these points on a chart, and measuring the true course and distance. Next the base weather station provided the weather enroute and the forecast winds. Luckily, unlike in the North Atlantic, the winds in the Pacific Ocean were mostly light and variable (essentially, no wind), at our unpressurized cruising altitudes of 9000 and 10,000 feet, depending on the direction of flight.

We were at Anderson Air Base (AB), Guam for refueling. I had completed my departure flight plan and was about to walk out to the aircraft. Just then, a C-141 crew burst through the entry door. The navigator sought me out and thought he was doing me a favor. He said, "Are you going to Mactan AB? The winds are 275/35." This meant the winds were coming from 275 degrees relative to true north and were 35 knots. My jaw dropped to the ground.

The C-141 navigator saw my facial expression and said, "You are a C-141?"

I replied, "No, we are a C-124."

Whereas the winds at our low altitude were light and variable, they were considerably different at higher altitudes. I had visions of redoing my flight plan, but knew these winds were not correct for our aircraft altitude.

Lockheed C-141 jet-powered airlifter
Ref: commons.wikimedia.org

Fuel Planning for the Mission
In Military Airlift Command (MAC), unlike SAC, the navigator was responsible for accomplishing the fuel planning, using the nomograms in the C-124 Fuel Planning Manual. The rollup of the fuel calculations included the sum of following:

> Empty Weight of our specific aircraft (as documented in the AF Form 781 for our specific aircraft)
>
> Weight of the crew members at 200 pounds per person
>
> Cargo weight
>
> Fuel weight required to fly the mission, as per my flight plan
>
> 1500 pounds of fuel for takeoff and climb
>
> Approach and Landing; 15 minutes of fuel
>
> 15 minutes holding-fuel at destination
>
> Fuel to fly to the alternate airfield, per my flight plan

Once I calculated the sum of the above weights, I subtracted the total from 212,000 pounds, which was the maximum allowable gross weight of the aircraft. This calculated number was called unidentified extra fuel. Essentially, every takeoff was a maximum gross-weight takeoff.

Takeoff from Wake Island
I recall our takeoff from Wake Island. It was hot. The pilot lined up the aircraft on the runway centerline and set the brakes. He then pushed the throttles to the limit and released the brakes. The aircraft slowly accelerated down the runway. My only outside view was through my left side window. I watched the runway flash by and then there was nothing but water. I looked at the altimeter, and it showed 50 feet and indicated the aircraft was climbing slowly. I

sucked in my stomach and muttered to myself, "Please don't lose an engine now. Please don't lose an engine now."

Anderson AB, Guam

We phased through Anderson AB in Guam just to refuel. Our departure was at night. Ordinarily, I hung close to the pilot in preparation to go out to the aircraft. I went to the bathroom, and when I came out, I was alone. I left the operations building into a pitch-black night. The only thing I could see were the dim red cockpit lights of our aircraft in the distance.

As I walked toward the lights, I heard a loud roar in the far distance. The noise seemed to last for a minute, and then white landing lights came on. The roar came towards our aircraft and as it passed by, it was obvious the aircraft was a B-52 taking off, heavy, for a bombing mission. A minute later, the scene was repeated. As each bomber made its takeoff, the sight was truly memorable. The grouping of B-52 bombers took off on a bombing mission in Vietnam, and from their sluggish performance on takeoff, they were obviously heavy.

B-52 Internet Photo
http://en.wikipedia.org/wiki/File:Boeing_B-52D-40-BW_%28SN_56-0695%29_and_GAM-72_Quail_decoy_missile_and_trailer_061127-F-1234S-010.jpg

Mactan AB, The Philippines

We landed at Mactan for crew rest and refueling. I recall waking up and walking to the officer's club for breakfast. The heat and humidity were oppressive. Also, the natives had defecated in their fields, to fertilize their crops, and the smell was overwhelming until the waste dried up.

We took local transportation into Cebu City, which was nearby, for a shopping trip. As a crew, we cleaned out a local shop of its monkey wood salad bowls and other merchandise.

Cam Rahn AB in South Vietnam

The last leg of our journey was into Cam Rahn AB in South Vietnam. This was a relatively safe base, as the landing approach was from the sea. When we landed, it was obvious this was a major cargo destination. The ramp was lit up with floodlights, and there were many charter airline aircraft represented, discharging and loading cargo.

We had another navigator on our trip. When we reported-in to the Airlift Command Post at Cam Rahn AB to file a flight plan outbound, he asked if there was any cargo going to "the front," wherever that was. Our pilot jumped all over him. "Are you crazy? This is a C-124, for God sakes!"

The airfield ramp was brightly lit with floodlights and there were aircraft all over the ramp loading and unloading cargo. Many aircraft were civilian airliners under contract to the government. I walked off the brightly lighted ramp to go to the men's room, and, when I was in the darkness, I nearly walked over a soldier standing guard with his weapon at the ready. I think I shocked him as well as myself. During the time we were on the ground, there was a C-123 aircraft flying circles all night long over the ground approaches to the base, inland. The aircraft dropped flares by parachute constantly, to illuminate the area and prevent the bad guys from sneaking up and making trouble

Engineer, Set Refueling Power

C-123K "Provider"
Ref: en. wikipedia.org

We discharged our cargo of two overhauled and fully assembled Hamilton Standard 54H60 propellers for the C-130 aircraft. This was the system I was responsible for at this company. Our load going outbound consisted of propeller blades from the C-123 aircraft headed for overhaul in the U.S.

We were prepared to depart, but had a problem with one engine magneto requiring a delay of a couple of hours to repair.

I took military leave from Hamilton Standard to go on this trip. Their policy relative to payment for military leave was as follows: They took the military base pay plus flight pay, the taxable portions of the pay package, and made up the difference relative to the company salary. Military pay consisted of base pay, flight pay, subsistence, and housing allowance. The latter two categories were tax exempt and the company didn't count them toward the pay differential calculation.

In those days, a person performing service in the Vietnam War zone was exempt from income taxes, for that month. When I returned to work, I submitted my paper work to the company on the basis I didn't earn any taxable income. Immediately, the company did a backflip, and said it would make the calculation on the basis of an assumption of taxable income. I thought it was interesting company policy didn't cover that situation, and Personnel had to come up with something quick to cover their ass.

The interesting thing about my trip to Vietnam was the cargo we hauled from Travis AFB to Vietnam consisted of three overhauled Hamilton Standard 54H60 built-up propellers for the C-130 aircraft. I worked for Hamilton Standard as an installation engineer and was responsible for this propeller system. On the return trip from Vietnam, we had crates of propeller blades from the C-123K aircraft, headed back for overhaul. In later years, I flew the C-123K as a navigator, but this is a story for another chapter.

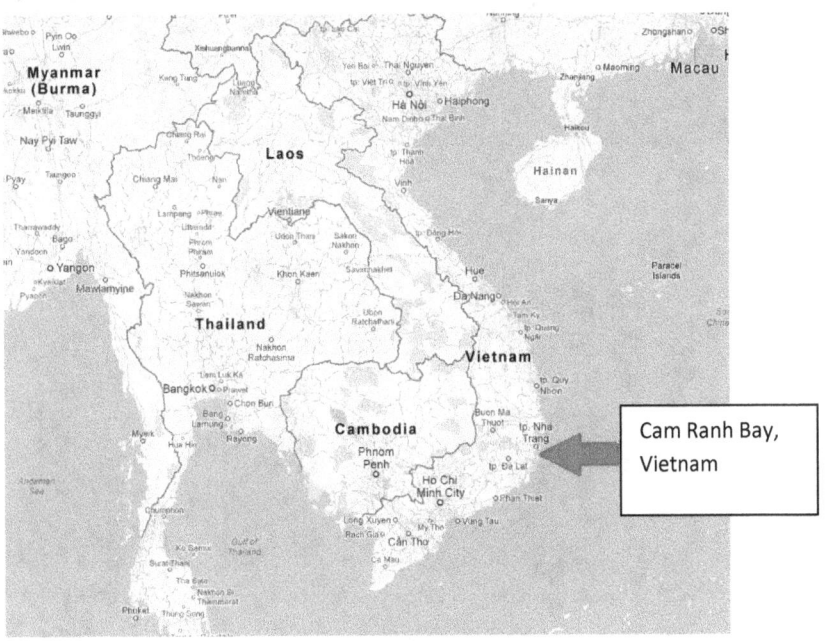

Cam Ranh AB in Vietnam
Ref: en.wikipedia.org

Pilot is on the left, copilot is on the right.
Photo by Author

Pilot Station
Note the repeater navigation radar scope to the left of the control column, and the nose wheel steering wheel to the left of the radar. Photo by Author

Synergy with My Civilian Job

As I stated, I worked as a propeller installation engineer at Hamilton Standard, and was assigned the 54H60 propeller system installed on the C-130 and P3 aircraft. I was intimately familiar with the hydramatic propeller system Hamilton Standard had in production for many years.

Now I was flying as a navigator on the C-124C aircraft, which was equipped with the Curtiss Electric propeller system. Curtiss was out of the propeller business for some years at this time. This large 17 foot diameter propeller utilized an electric motor in the hub with a reduction gear train to change the propeller pitch. For steady state conditions, there was a powerful clutch preventing the pitch from changing. Also, the blades were of hollow steel construction. Essentially, there was a central load carrying spar, with foam surrounding the spar to form the airfoil. The outer sheath was a layer of steel.

Number 2 Engine and Propeller on the C-124C Globemaster II
Photo by Author

Note the individual engine exhaust stacks. The aircraft was unpressurized, and therefore limited to 10,000 feet altitude unless the crew was on oxygen. There were no turbo superchargers in the engine exhaust. There was, however, a gear driven supercharger, with two stages. The flight engineer shifted from low to high blower during climb.

This photo was posted on our Flight Safety Board. The number 4 engine had a massive internal failure and seized up before the propeller could be fully feathered

Number 1 and 2 Engines.
Photo by Author

It Rained Inside the Aircraft

As the aircraft was unpressurized, whenever it flew through precipitation, the water found its way through minute voids in the sheet metal construction of the fuselage. Water dripped from the ceiling over everything; the crew, equipment, etc. It was bazaar.

Cross-Country to Roosevelt Rhodes Naval Station

I signed-up for a three-day trip to Roosevelt Rhodes Naval Station in Puerto Rico, Friday through Sunday, 23 August 1968 returning 25 August 1968. There were three aircraft scheduled for the cross-country. We were all on AFTP status, which meant we had to average 2.5 hours of flying time each day, over a six month period, in order to be paid.

During the flight briefing, we received instructions in the event of an aircraft abort. If we didn't fly, we didn't get paid. In the event of an abort, one of the other two aircraft was to return to pick-up the crew members on the ground.

I was on the third and last aircraft scheduled to depart this evening. Sure enough, our aircraft was not flyable. Neither of the other two aircraft returned to pick us up. We hung around all evening while the maintenance organization prepared another aircraft for flight. Finally, about 0100 hours Saturday, we were able to takeoff.

After an eight hour flight to Roosevelt Rhodes, we landed about 0900 on the 24th. I was completely wrung out without any sleep all night. I now had a decision to make. The other two crews were lined up to take the U.S. Navy bus to San Juan, as we were on the opposite side of the island. Should I stay on base and cool it around a pool, or should I take the bus? I elected to do the latter, which turned out to be a major mistake.

It was a long ride to San Juan, and the driver let us off quite a distance from the tourist section of the city. It was August, and it was hot and humid. We started walking. And we walked and we walked and we walked. It seemed we walked forever. By the time we arrived in the built-up section of San Juan, I was completely sweat-soaked and exhausted from lack of sleep.

I do not remember what I did all day, but, midafternoon, I found an air conditioned hotel lounge and I camped out until the appointed time for pickup by the U.S. Navy bus. I cannot remember why we didn't ride the bus to this point where we were picked up later.

After arrival back at Roosevelt Rhodes, I crashed and caught minimal sleep before being awakened early to flight plan for the return leg. This was a tough way to earn a few extra bucks. I did not plan to do this again. It definitely was not fun.

Wing Spar Cracks

In my last year with the 337th MAS in the Air Force Reserve, wing fatigue cracks were discovered in the majority of the remaining C-124 aircraft in service. Since the C5A turbofan-powered airlifter was just coming into the inventory, the Air Force intended to keep the reserve forces C-124 aircraft just long enough for the more modern aircraft to become operational in sufficient numbers. It was not economical to repair the wing cracks in the older C-124 aircraft.

The Air Force performed a one-time inspection of the wing spars on all C-124C aircraft, and retired those with a potential for failure. Then they placed a restriction on the remaining C-124C aircraft, in order to minimize the fatigue stress on the wings. The cruise airspeed was limited to 200 knots true airspeed (KTAS), and we aircrews were precluded from filing a flight plan into areas of known or forecast moderate turbulence. Also, we were limited to one takeoff and landing per flight. These restrictions were intended to minimize the fatigue stress on the aircraft wings.

Engineer, Set Refueling Power

C-124C taxiing.
Hear the "Chugga, Chugga, Chugga?"
Ref: commons.wikimedia.org

Here we were, returning to Dover AFB from an airlift mission to Lajes AB, in The Azores. As we approached the Air Defense Identification Zone (ADIZ), I attempted to get as good a position as possible. We had to hit this zone within 5 minutes of my forecast, or risk being intercepted by a fighter interceptor to find out who we were. Our aircraft was in the "soup" at our cruising altitude of 10,000 feet, so celestial navigation was not a consideration. My best navigation aids were Loran and pressure pattern.

The aircraft picked up a light "chop." The turbulence got worse and worse. Now I was definitely becoming concerned. My Loran scope shook so hard on its shock absorber mounts, I couldn't read the traces. We had another navigator on-board who had over 3000 hours in the C-124 on active duty. He volunteered to take my seat and get a fix. He wrapped his arms around the scope to steady it, but could not get a reading. He looked across the aisle at the flight engineer, who was throwing up. Next my substitute navigator developed dry heaves.

I got back into the navigator seat. The copilot called me on interphone and told me he could see some stars and I should get my sextant out. We must have been flying along the top of cumulus clouds, and were in and out of the soup. This was ridiculous, as the aircraft needs to be stable without significant turbulence in order to get a reasonable celestial navigation fix. I was seriously concerned the aircraft would break up due to a wing structural failure. I knew too much about fatigue stress from my engineering background. At this point, I was ready to trade my wings for a spot on dry land. I was scared! I had visions of the aircraft breaking apart and I was floating in a life raft, hoping someone would spot me and any other survivors.

We plodded along, as the aircraft made monotonous noises, Chugga, Chugga, Chugga, and eventually made our way into Dover AFB. Immediately after landing, we marched into the squadron operations building, into the weather station and told the meteorologist about the turbulence issue. He sent a "Green Flash" message to all locations in Europe to hold all C-124 aircraft until further notice.

North Korea Captured the USS Pueblo and its Crew
The USS Pueblo, a U.S. Navy intelligence gathering (aka spy) ship, was attacked and seized by the North Korean military on January 23, 1968, just off the coast of North Korea.

The following is quoted from en.wikipedia.org

"The seizure of the U.S. Navy ship and its 83 crew members, one of whom was killed in the attack, came less than a week after President Lyndon B. Johnson's State of the Union address to the United States Congress, just a week before the start of the Tet Offensive in South Vietnam during the Vietnam War, and only three days after 31 men of North Korea's KPA Unit 12 had crossed the Korean Demilitarized Zone (DMZ) and killed 26 South Koreans in an attempt to attack the South Korean Blue House (executive mansion) in the capital Seoul. The taking of *Pueblo* and the abuse and torture of its crew during the subsequent 11-month prisoner drama became a major Cold War incident, raising tensions between the western democracies and the Soviet Union and China.

North Korea stated *Pueblo* deliberately entered their territorial waters 7.6 nautical miles (14 km) away from Ryo Island, and the logbook shows that they intruded several times. However, the United States maintains the vessel was in international waters at the time of the incident and any purported evidence supplied by North Korea to support its statements was fabricated.

Pueblo, still held by North Korea today, officially remains a commissioned vessel of the United States Navy. Since early 2013, the ship has been moored along the Potong River in Pyongyang, and used there as a museum ship at the Pyongyang Victorious War Museum for propaganda purposes. *Pueblo* is the only ship of the U.S. Navy still on the commissioned roster currently being held captive."

While employed by Hamilton Standard as a propeller installation engineer I returned from a business trip to the west coast. While changing airplanes at Chicago's Midway Airport, I passed a newsstand and glanced at the headlines. It stated President Johnson called up the Air Force Reserve in response to the Pueblo Crisis. Holy cow!

I found a phone booth and quickly called Marilyn at home. I asked her if I received a recall notification from my reserve squadron. She replied, "No." There was a selective call-up of the Air Force Reserve and Air National Guard, but my unit was not among those called up.

Two years later, I was serving in the Ohio Air National Guard in their KC-97L air refueling tankers. A peer lieutenant navigator had been in the Georgia Air National Guard at the time of the Pueblo incident, and his C-124C Globemaster II unit was recalled for a year of active duty. I would not have minded being recalled, but it didn't happen.

Ref: en.wikipedia.org; By Laika ac from USA - USS Pueblo, CC BY-SA 2.0,
https://commons.wikimedia.org/w/index.php?curid=31723345

The C-124 was Nicknamed "Old Shaky"

The C-124 aircraft was affectionately called "Old Shaky." It vibrated its way through flight. One engine was identified as the "master" and the engine speed, (rpm or revolutions per minute) of the other three were synchronized to it, by adjusting the pitch, temporarily, on the propellers. Old Shaky still vibrated. It was as if there was a standing vibration wave with another superimposed to reinforce and attenuate the main wave. We applied other names to the aircraft, such as "The Aluminum Overcast" and "50,000 rivets flying in formation."

My First Experience with Hearing Attenuation

We returned from Lajes AB in The Azores to our base at Westover AFB, Massachusetts. After completing the paperwork to get paid, I entered my old Volkswagen bus to drive home. Suddenly I realized my hearing was attenuated. The bus made the air-cooled engine noise of "Clatter, Clatter, Clatter," but I barely heard it. The next day my hearing returned. This was the beginning of my awareness a hearing issue was developing.

C-124 Globemaster II
Ref: nationalmuseum.mil

Easy Weekend Trip to Goose Bay, Canada

I signed-up for a weekend three-day trip to Goose Bay AB, Canada. This should have been a piece of cake. We took off after work on a Friday evening to fly to Dover AFB, Delaware. There, we were loaded with two pallets of mail headed to Goose Bay AB, Labrador. There had not been an aircraft to Goose for two weeks, so mail was important to the isolated tour personnel there.

As we flew north from Dover, I had visions of landing at a reasonable hour and eating a steak supper at the officer's club. We landed after 2230 hours local time (10:30 P.M.). The base mess hall was closed as was the officer's club. We ate our supper in the office of the BOQ; a beer and a candy bar. I was pissed. The Air Force simply didn't have its priorities straight. All the base functions were set-up for the convenience of the full-time personnel. I suggested to our pilot we put the two pallets of mail back on board the aircraft, but he would not hear of it. This was just the way it was.

Weekend Trip to Bermuda

I was the navigator on a weekend active duty trip to Bermuda. Once this was Kindley Air Force Base, but at some point, the Air Force pulled out, and it was assigned to the Navy, as Naval Air Station Bermuda.

We departed on Saturday morning for an easy flight to Bermuda. This was another easy weekend flight, in daylight, to Bermuda. We were at our cruising altitude of 9000 feet, the aircraft was trimmed for autopilot operation, and the pilot slept in the bunk behind my station. This was allowed on long range flights, as only one pilot had to be in the "seat."

My instructions were to install my periscopic sextant when I needed to use it; otherwise I was supposed to remove it. The reason was it hung down about 12 inches from the mount on the aircraft ceiling, and it was possible to bang your head walking by. I disregarded this protocol, and left the sextant in its mount.

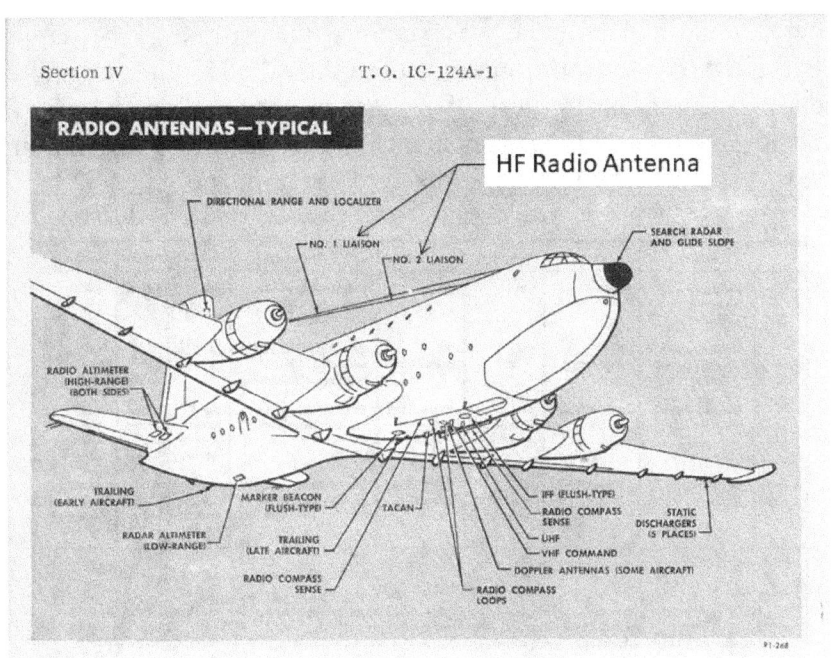

Location of pair of HF Radio Antenna
Ref: C-124A Flight Manual

We were in the clear, with a few cumulus cloud build-ups about 30 nautical miles off to our left. I saw them on my radar, but was not concerned. Everything was quiet in the cockpit, except for the monotonous engine noise, "Chugga, Chugga, Chugga." Suddenly, without warning, there was a tremendous flash of light for an instant, followed by a **CRASH.**

Engineer, Set Refueling Power

Our sleeping pilot came awake from his stupor, clawed his way past my sextant, while banging his head on it, and dropped into his left seat. The engines continued their, "Chugga, Chugga, Chugga."

We had been hit by lightning. Immediately, my radar and Loran failed. I examined the top of the aircraft through the sextant and saw there was only a short length of one of the HF antenna left, fluttering in the slipstream. Normally, this antenna pair ran from the top of the cockpit to the tail of the aircraft.

After landing at Bermuda, we examined the aircraft. The damage was limited to a few burn marks on the skin, the HF antenna and my radar and Loran electronics. We changed our shorts and had a fun time in Bermuda. The maintenance personnel made the required repairs to the aircraft in time for us to return home on schedule.

C-124C at the Air Force Museum, with the Clamshell Doors Open and Loading
Ramps Deployed
Open Wide and Say "AAHH"

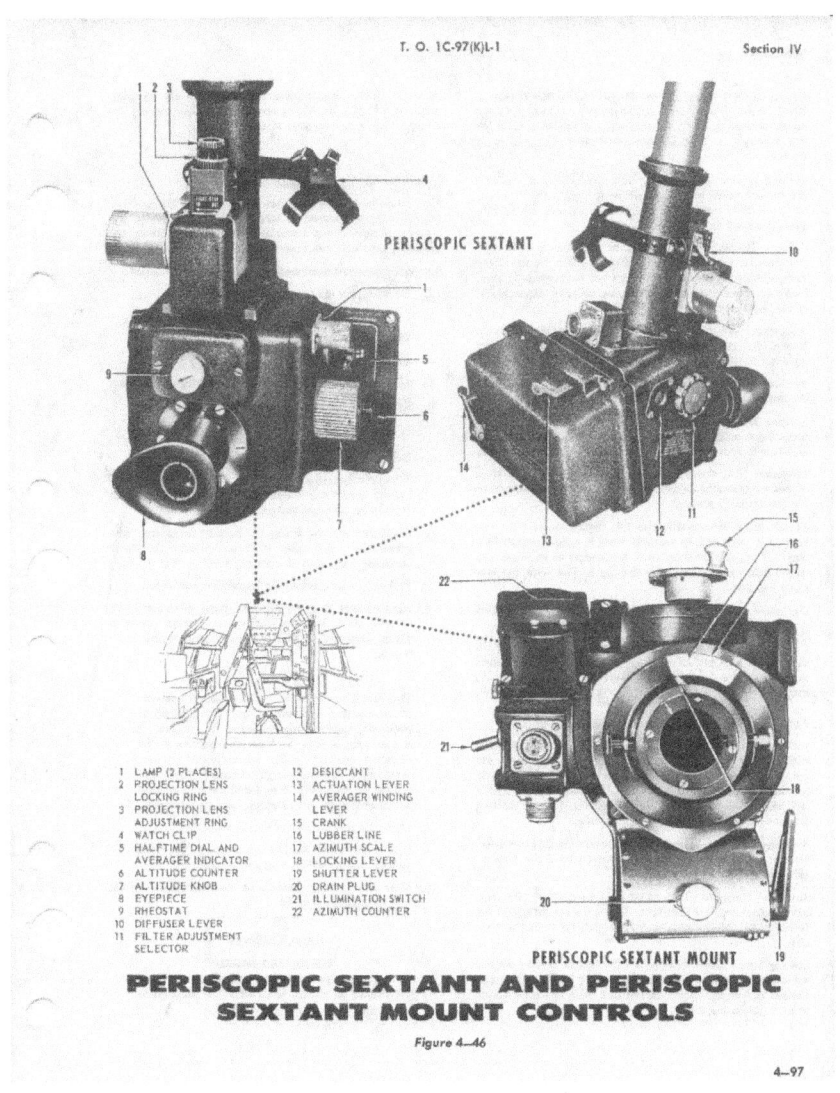

PERISCOPIC SEXTANT

1	LAMP (2 PLACES)	12	DESICCANT
2	PROJECTION LENS LOCKING RING	13	ACTUATION LEVER
3	PROJECTION LENS ADJUSTMENT RING	14	AVERAGER WINDING LEVER
4	WATCH CLIP	15	CRANK
5	HALFTIME DIAL AND AVERAGER INDICATOR	16	LUBBER LINE
		17	AZIMUTH SCALE
6	ALTITUDE COUNTER	18	LOCKING LEVER
7	ALTITUDE KNOB	19	SHUTTER LEVER
8	EYEPIECE	20	DRAIN PLUG
9	RHEOSTAT	21	ILLUMINATION SWITCH
10	DIFFUSER LEVER	22	AZIMUTH COUNTER
11	FILTER ADJUSTMENT SELECTOR		

PERISCOPIC SEXTANT MOUNT

PERISCOPIC SEXTANT AND PERISCOPIC SEXTANT MOUNT CONTROLS

Figure 4-46

4-97

Ref: KC-97 Flight Manual

Driftmeter Behind Sheet Metal
Protective Cover

Photo by Author

The driftmeter in the C-124 aircraft was located on the main cargo deck. This necessitated leaving the cockpit, and climbing down the 20 ft. ladder to make a reading. The driftmeter was protected by a sheet metal cover from inadvertent contact with cargo.

Crews Named the Aircraft "*Old Shaky*"

Aircrews sometimes name their aircraft. For example, the WWII P-47 Thunderbolt was named the *jug* by the aircrews, probably referring to the finned air-cooled engine, equipped with cylinders referred to as jugs. The B-52 heavy bomber was named by its aircrews as the *buff*, or big ugly fat fellow; the last word was really a nasty word starting with the letter *f*.

The C-124 was equipped with an engine/propeller synchronizer, which matched each engine rotational speed to a selected *master* engine to minimize engine vibration. However, the technology in the 1950s hadn't yet advanced to match the phase angle of each propeller blade to that of the others, which was more effective in reducing engine vibration. Accordingly, there was a pulsating vibration constantly throughout the airframe. It would wane and then strengthen. Accordingly, the crews called the aircraft, *"Old Shaky."*

Monitoring Cumulus Buildup Ahead of Us

One day, we were heading somewhere, I forget where. The pilot called me on interphone and inquired why I wasn't vectoring the aircraft around some heavy cumulous cloud buildups ahead of us, using my search radar. I told him in no uncertain terms, if there were buildups in front of us, he needed to tell me about them, as I had no forward vision due to my equipment and bulkhead in front of me. I could not stare at the radar sweep in front of me without becoming hypnotized. He might be the first at the scene of an accident, but I would be right behind him!

Runway at Lajes AB, The Azores

Lajes AB, Terceira, Azores, Internet photo
Ref; militarybases.com

There was virtually no level ground on the base at Lajes AB. One either walked uphill or downhill, everywhere. I flew there many times while associated with the C-124 aircraft. We were advised when we checked into the visiting officer's quarters (VOQ) we could put our shoes outside the door, and the shoe shine boy (a man old enough to be my father) shined them overnight. We were cautioned to pay him no more than a nickel, because the mayor in town didn't want the shoe shine boy making more money than the mayor.

On one trip into Lajes AB, we passed an ocean station vessel (a ship anchored at a navigation point to provide navigation assistance to aircraft flying overhead). Normally, I requested a fix from the ocean station vessel to help me, but this time the ship had been in the soup for two weeks, and didn't have a good position from which to provide me a fix on our position. It was Mother's Day, and our pilot took radio messages from the ship's crew to forward to the crew's loved ones when we landed. Upon landing at Lajes, our pilot bought Mother's Day cards with his own funds, wrote out the messages and mailed them. I thought this was a touching thing to do. Over the next three years, I flew around the world with this unit.

Summation of Flying in a C-124

During the three years I was associated with this aircraft, I traveled many places; Newfoundland, Bermuda, England, Iceland, Germany, Spain, Puerto Rico, Hawaii, Wake Island, Guam, Philippines, Vietnam, all either on three-day weekend active duty tours, 15 day active duty tours, or AFTPs. The C-124C cruised at 200 KTAS (Nautical Miles per Hour, True Air Speed) at 9000 to 10,000 feet altitude, depending on the direction of flight, unpressurized. The limitations on speed, fuel load, and cargo meant the aircraft needed to island hop in order to go long distance. It seems no matter how well we planned, we started out flying in the day, and sleeping at night, but we quickly ended up flying all night and sleeping all day. When it rained, due to the lack of aircraft pressurization, the rain leaked into the aircraft skin, and dripped over everything, including me.

Chapter 9; OHIO AIR NATIONAL GUARD AND KC-97L EXPERIENCE

I left the employ of Hamilton Standard in November 1969, after four years, and joined GE Aircraft Engines in Cincinnati, Ohio. I also resigned my commission in the Air Force Reserve to accept a commission in the Ohio Air National Guard.

Transition from Air Force Reserve to Ohio Air National Guard

I had a problem in transitioning from the Air Force Reserve to the Ohio Air National Guard. The Air Reserve Technician with the C-124 equipped 337th MAS wanted to hold onto me, on paper, as my departure would have impacted the unit's C1 status (combat ready number of crews). He made a deal with the ANG Air Technician to have me train with his unit in Ohio until a replacement navigator could be qualified at Westover AFB.

Presumably, I would train with the ANG and be paid by the Air Force Reserve. I foolishly acquiesced to this deal, and proceeded to attend weekend drills and fly AFTPs with the Ohio Air National Guard.

Unfortunately, I earned over $500 doing this, and I hadn't been paid. When I called my unit at Westover to talk to the Air Reserve Technician, I learned he had been removed from flying status, due to a medical disability, and was no longer the contact. Nobody else knew about the "Deal." It turns out it was not legal to be paid by the Air Force Reserve for duty with the Air National Guard! It took letters to Ohio Senator Robert Taft to get it straightened out, but eventually it was. I was paid, and I received my commission in the Ohio Air National Guard as a captain.

History of the ANG Conversion of the KC-97G to the KC-97L

The Illinois ANG was originally equipped with the KC-97G Tanker for use in refueling their fighter aircraft; these tanker aircraft were released from the SAC inventory. When the Illinois ANG learned of USAF plans to phase out its KB-50 tankers, and of the airplane's structural commonality with the KC-97 due to the use of identical wings, they suggested using the KB-50 auxiliary jet engines on their KC-97G models. This was approved at the appropriate level of command, and direction was given to modify one KC-97G.

An in-flight evaluation demonstrated the capability to refuel the F-84F, F-100, F-101, B-66, F-4C and F-105 aircraft, at altitudes above 20,000 ft., and at airspeeds compatible with these receiver aircraft. The reduced piston engine power demand resulted in reduced spark plug changes; there were only five spark plug changes required during the evaluation compared to 50-60 changes that would have been necessary during the evaluation for the KC-97G model.

The evaluation demonstrated the improved time to climb to 20,000 ft. altitude. It took only 22 minutes for the "L" model vs. 48 minutes for the "G" model on the piston engines alone. (Author's comment; the time to climb for the KC-97G was undoubtedly to 15,000 ft. altitude, as it is hard to believe the KC-97G could climb to 20,000 ft. at comparable weight for the KC-97L in 48 minutes.)

My First Flight in the KC-97L
My first flight in the KC-97L with four reciprocating engines and two jets was exhilarating. Upon brake release, the aircraft literally leaped forward, broke ground and climbed to 25,000 ft. altitude in 12 minutes flat (Reduced climb time vs. that specified above was due to differences in aircraft gross weight for the test, undoubtedly.) The flight engineer, who had previous experience with SAC KC-97s, told me on his first flight in the "L" model, the aircraft passed through 10,000 feet so fast he almost forgot to start the cabin pressurization. Let me try to provide a comparison between my experience in the SAC KC-97G with just the piston engines, and the KC-97L in the OANG, with both the four piston engines and the two jet engines. See the next page.

As the aircraft shot down the runway, the flight engineer stated, "Engineer Adjusting Power." This was necessary as the pilot's eyes were outside the aircraft, not on the engine instruments, and the flight engineer had to keep from over boosting the engines on takeoff. I had to hold everything loose on my navigator desk from sliding aft and onto the floor. I never got over this feeling of throbbing raw power! It is difficult to explain, but it was almost like a drug high.

Performance Comparison of KC-97G vs, KC-97L

Aircraft	Engines	Time to Climb	Cruise Altitude	Cruise Airspeed
KC-97G	(4) R-4360-59B	45 minutes.	15,000 ft.	220 KTAS

(Aircraft Gross Weight of 175,000 lbs.)

Aircraft	Engines	Time to Climb	Cruise Altitude	Cruise Airspeed
KC-97L	(4) R-4360-59B	12 minutes	25,000 ft.	275 KTAS
	(2) J-47			

(Aircraft Gross Weight of 153,000 lbs.)

Acceleration Check Times

KC-97G	10 – 15 seconds,	60 to 100 KIAS speed on the takeoff
KC-97L	5.0 seconds or less.	60 to 100 KIAS speed on the takeoff

The ANG also added rendezvous radar to the KC-97L, the AN/APX-29A, which allowed the navigator to actually direct the rendezvous with fighter aircraft not equipped with radar, such as the North American F-100 Super Sabre and the Cessna A-37 attack aircraft. The radar operated by triggering the fighter Mode 2 on their IFF/SIF equipment, and displayed this coded return signal on the rendezvous radarscope. Also, the aircraft were modified with an altitude-hold feature on the autopilot. This greatly improved the accuracy of celestial navigation, by improving the stability of the aircraft.

The KC-97L aircraft assigned to the 145th Air Refueling Squadron were former SAC aircraft having spent a good part of their lives sitting on alert, but not having a lot of flying hours on the airframe. Consequently, the airframes were in good condition for their age. The addition of the two J-47 jet engines and the lighter gross weight operation resulted in a much more reliable airframe and piston engines than we had in SAC. It also helped the aircraft were in the ANG, which had low personnel turnover in both maintenance and aircrews compared to an equivalent active duty squadron. The reason for this was the active duty squadrons had transfers, expiration of service commitments, and schooling assignments. As a result, the active duty force was in a constant training mode compared to the ANG, which had a relatively stable personnel base.

The refueling track used was the same one used by SAC when I was on active duty; it ran from east to west across upstate New York State, across the Finger Lakes, and ended just before Erie, Pennsylvania. Rendezvous and refueling was at 25,000 ft. altitude and 250 to 275 KTAS, in level flight. The aircraft performance was such the tanker could actually accelerate ahead of the Cessna A-37 aircraft (but we could not accelerate ahead of the faster F-100 aircraft.)

Operation Creek Party

The Air National Guard tanker units were tasked with a full-time presence at Rhein-main AB in Germany (co-located at Frankfurt International Airport). This was termed "Operation Creek Party," and was established to provide ANG tanker air refueling support for the NATO fighter aircraft in Europe. The active duty KC-135 jet tankers were tied up both with their SAC alert responsibilities and in support of the Vietnam War. This was the way we fulfilled our required annual 15-day active duty tour, a requirement to stay in the active Air National Guard. Every two weeks a different KC-97L unit rotated in from the United States to Rhein-main AB. Operation Creek Party was initiated June 2, 1967 and ran for ten years. Among the NATO fighter aircraft refueled were the F4D and F4E Phantom fighter, although all NATO aircraft received air-refueling support.

Transitioning of the KC-97 Air Refueling Tanker from SAC into the ANG

When I transitioned from the KC-97G on active duty in SAC to the KC-97L in the Ohio Air National Guard (OANG), it was like transitioning from a powerful but heavily loaded truck to a hopped up V-8 powered Corvette! The ANG was the recipient of some of the retired KC-97G tankers from SAC. The ANG modified the aircraft by retrofitting two GE J-47 booster turbojet engines under the wings for better aircraft performance. The KC-97L could take off in a shorter field length, climb faster, cruise faster and higher. The air refueling could take place at a higher altitude and at a higher speed.

Pilots in the 145th Air Refueling Squadron

In most AFRES and ANG flying units, the majority of the pilots had full-time jobs flying for an airline. It figured, as most units were located near a big city, where a major airline hub existed. These airline professionals maximized their allowable flying time for the month in the first two weeks. Then they flew with the reserve forces for the balance of the month. These pilots hung around together, and looked down upon the navigators. I believe the rationale for this attitude dates back to World War II. Many, perhaps most, pilot candidates did not make the grade. After "washing out," many of these men subsequently graduated from either the navigator or bombardier program. Pilots looked down upon these crew members as they didn't make the grade as a pilot.

When I was in the OANG, the pilot cadre didn't fit this mold. The unit was not initially located near a big city, but rather was in Wilmington, Ohio. When the government closed Clinton County Air Reserve Base, the AFRES and ANG units moved to Lockbourne AFB, near Columbus, Ohio. However, the turnover in aircrews was low, so it took a long time for airline pilots to backfill. In my unit, it seemed the navigators were "king of the hill." The pilots I remembered were rural postmen, small grocery store proprietors, etc. They acted the equal of the navigators, and it was refreshing.

Refueling Missions Revolved Around Time

The takeoff was predicated on the planned enroute time to complete the rendezvous with the receiver aircraft at the Air Refueling Control Point (ARCP). An orbit of 15 minutes was factored into the mission. The navigator obtained a time hack from the National Bureau of

Standards Radio Station, WWV, a time based on the atomic clock. The station announced the time every five minutes. Even better, the Canadian station, CHU, announced the time every minute.

Once at the aircraft, and after preflight was accomplished, the navigator kept the crew updated on the time to engine start. After the one minute call, I made calls over interphone at 30 seconds, 15, and then 10 down to zero, or hack. At precisely engine start time, the flight engineer pressed the starter button for number three engine. If this was to be a cell, or formation, an observer witnessed the number three propeller start to turn at exactly the same time, for each aircraft.

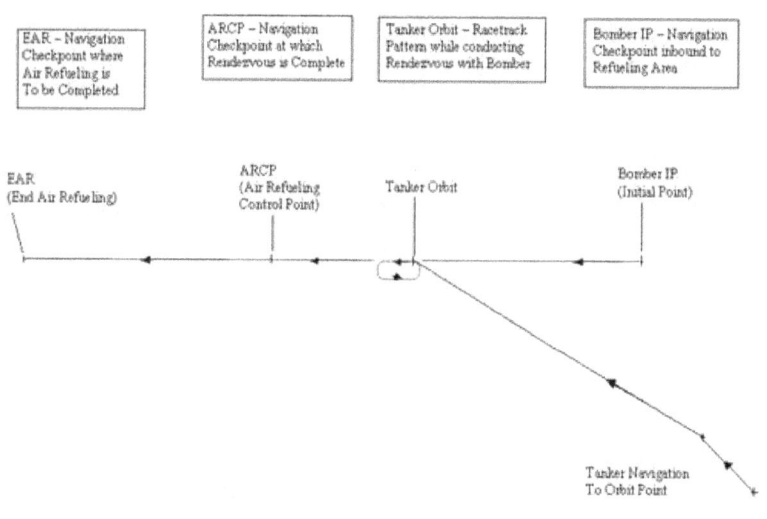

Overview of a typical Air Refueling Track
This art was generated for bomber refueling, but applies to fighters as well.

Engineer, Set Refueling Power

ANG KC-97L

Ref: national.museum.af.mil

The ANG KC-97L had
GE J-47 Turbojet Engines
On the Wing Pylon

The SAC KC-97G had
External Fuel Tanks
On the Wing Pylon

SAC KC-97G

Ref: commons.wikimedia.org

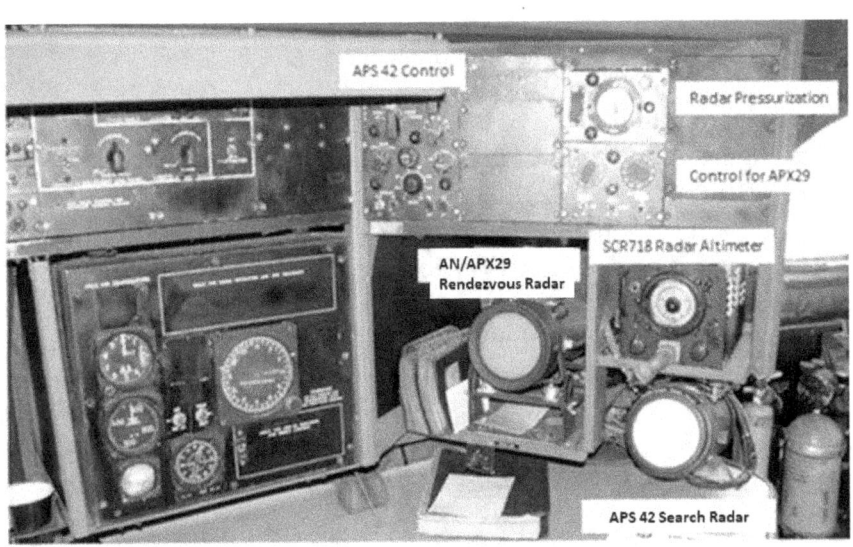

KC-97L Navigator Search Radar and Rendezvous Radar
Photo by Author

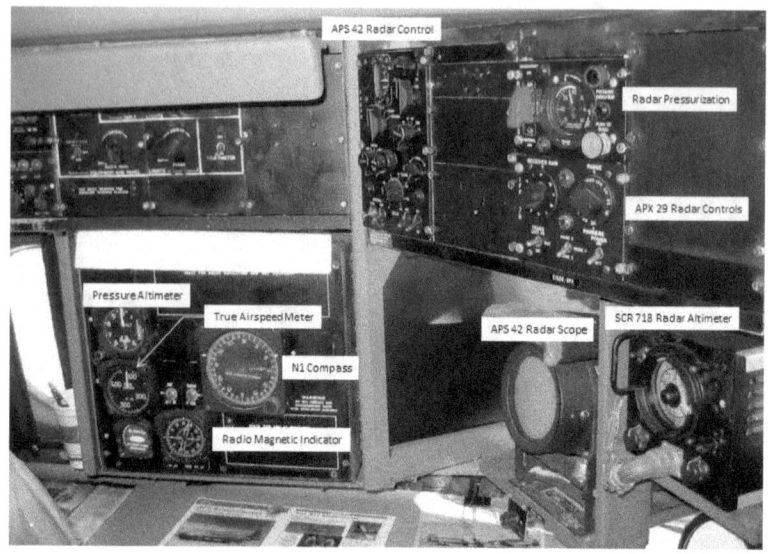

KC-97L Navigation Instrumentation
Photo by Author

The navigator similarly kept the crew updated on the time-to-takeoff. After engine start, the pilot taxied the aircraft out of the parking spot at the 30 minutes-to-takeoff call. Every takeoff was a "controlled takeoff." After engine warmup, checkout and propeller checkout, the pilot requested permission to take the active runway following my five minute call and he then applied the brakes. The flight engineer set the throttles to bring the piston engines up to *barometric* intake manifold pressure. (The gear driven supercharger crammed air into the carburetor to the level of atmospheric pressure.) When the flight engineer stated over interphone, "Power Stabilized, Sir," the pilot stated over interphone, "Pilot's Throttles" and set the throttles for maximum power. The copilot toggled the two switches controlling the jets (the throttles for the two jets were spring loaded switches) to 100% speed; the aircraft just **strained** against the brakes.

There was a massive throbbing throughout the airframe, and the aircraft just itched to move. The pilot released the brakes, under my countdown to the scheduled takeoff time, and the aircraft leaped forward. As it accelerated down the runway, the flight engineer stated, "Engineer Adjusting Power." This was necessary as the pilot's eyes were outside the aircraft, not on the engine instruments, and the flight engineer had to keep from over-boosting the engines with their gear-driven superchargers, on takeoff. I had to hold everything loose on my navigator desk from sliding aft and onto the floor.

Pilot's throttles and flight engineer's throttles were interconnected
Photo by Author

Immediately after takeoff, I directed the aircraft using the search radar and radio aids. I never got over the feeling of throbbing raw power! It is difficult to explain, but it sent shivers throughout my body.

Good Show, Nav

The wing commander flew regularly on our training missions, as pilot in command (aircraft commander) as well as on our deployments to Germany. I reported-in to fly one Friday afternoon.

Engineer, Set Refueling Power

Upon checking the scheduling board, I observed the wing commander was the aircraft commander on my crew. I made the mental observation, *"No relaxation tonight."*

We took off and climbed out. I navigated to the refueling area using radio navigation. This meant I was not in command of the compass headings. We flew radio facility to radio facility, just as airliners do today. A consequence was I was not able to calculate an inflight wind; I used the flight plan initial headings between radio facilities, which were based on the forecast winds.

I established the aircraft in the Orbit, and conducted the rendezvous with a pair of F-100 fighter aircraft from Mansfield Airport (Also OANG). The refueling was routine. After the refueling was completed, and the flight engineer brought the engines back to cruise power, The AC called me on interphone, "Hey Nav, Good show." I didn't know what he was referring to until after we landed and the copilot told me what happened. The aircraft hit the Air Refueling Control Point (ARCP) precisely on scheduled time and on the exact radio fix off Rochester TACAN. The AC kept pointing at the navigation instrument and saying, "Look at that, look at that." The amazing thing was it was all a coincidence, as I had not obtained an in-flight wind. The forecasted winds at altitude in this area were spot-on.

The instrument (TACAN) looks like an automobile odometer, counting off the miles from the selected radio station. I used preflight winds as I had not had time to obtain an in-flight wind, and the actual winds were exactly as briefed (an unusual occurrence). Anyway, from that time onward, every time I reported on Friday afternoon to fly, I checked the schedule; and noted I was flying with the colonel as a pilot. And it was all a coincidence! Try as I would, I was never able to repeat that experience hitting the ARCP exactly on time/target, but as far as the colonel was concerned, I could do no wrong.

Crew Briefing for "Creek Party" Deployment

I volunteered for a 15-day active duty tour to Rhein-main AB, Germany. There were six aircraft deploying at staggered takeoff times. The earliest takeoff time was in the morning. Our takeoff time was 1530 hours, mid-afternoon. The wing commander directed all the crews report for a briefing at 1030 hours (10:30 A.M.) I was not happy with this early report time, but complied. The briefing consisted of the following: "You all know what to do, so just do it. You know what to do, so just do it." I came in early just to hear that? It took perhaps three minutes. Now I had much of the remaining day with nothing to do.

R&R on the Middle Weekend

During the middle weekend of the two week tour, other ANG units scheduled an R&R flight to Majorca. Our wing commander refused to allow his unit this luxury. His point was, if the aircraft broke down in Majorca, he couldn't cover the scheduled active duty refueling on Monday. He probably had a point, but the crews were unhappy.

On the deployment, on the ground at Goose Bay AB, during the aircraft refueling, I stayed on board to review the overwater routing. I planned where I was to get fixes on our position going across the Atlantic Ocean. The lower forward entry door was open during the aircraft refueling operation, and the air got colder by the minute. I didn't notice it at the time, but my winter thermal underwear jacket had jacked up exposing my back to the cold air. After takeoff, and enroute, I kept trying to stretch while standing up, not realizing I got a cold in my back. By the time we got to Rhein-main AB, I was in agony with back spasms. This was the beginning of my problems with my back, which I have had ever since. I went to the flight surgeon and was grounded for much of the two weeks I was at Rhein-main AB.

Yo-Yo Missions

The air traffic in Europe in this timeframe (1970s) was congested. My AN/APX-29A rendezvous radar picked up only IFF/SIF Mode 2 signals (the channel for military aircraft, only). In the U. S., only the intended fighter aircraft signal was usually picked up, due to the scarcity of military aircraft in the air at any time. But in Europe, the Mode 2 signal on the scope was overwhelming, due to the large amount of military traffic. The airways congestion was incredible, also, due to both heavy commercial and military traffic. This, plus the real possibility of overflying and breaching sovereign country borders, and the relatively high speed of the refueling, precluded Air Traffic Control (ATC) from allowing the tanker navigator to conduct the rendezvous.

ATC managed the entire flight, including the rendezvous and refueling. I, as the navigator, was along for the ride. Missions were short. We referred to them as *Yo-Yo* missions. We took off, climbed out, were directed to the refueling track, the fighter was vectored in behind us, the refueling took place, and we descended and landed. Missions were as short as 1-½ hours. We might fly two or three times a week, with the rest of the time off!

Unexpected Heavy Offload for a Pair of F4 Phantoms
On September 1, 1971, we were assigned a refueling on Baumholder Air Refueling Track. As we were directed to the Orbit, the two F4E Phantoms were vectored-in right behind us by Air Traffic Control. We were scheduled to give each 5000 lbs. of fuel, but the fighters were low on fuel, and requested 15,000 and 16,000 lbs., respectively. They were right at *bingo* fuel, which meant if we were not there for the refueling, they would have had to make an emergency descent (called a jet penetration) and landing for fuel.

Aircraft Stacked Up
My crew flying a *Yo-Yo* mission out of Rhein-main AB, and had just finished the refueling of an F4D Phantom fighter, when we received clearance to descend (aircraft were stacked up at minimum altitude separation intervals, due to the high aircraft traffic, both military and commercial traffic.) We had on-board the air technician navigator. The pilot took his time running the "Descent Checklist," before actually descending, when the air tech jumped all over him. "Point

the nose down <u>now</u>, and <u>then</u> start running the Descent Checklist, unless you want the next higher aircraft in the stack coming down right on top of you." The air traffic was incredible in Germany at the time.

German National Private Pilot
While descending from a *Yo-Yo* mission, we listened on Ultra High Frequency (UHF) Radio to the air traffic controller talking to several aircraft at the same time, as was his practice. One of the aircraft apparently was a light airplane flown by a German national. The controller requested the light aircraft position, and was given the same. The controller immediately said, "You are ***Wrong***, you are not where you say you are." The controller then contacted several other aircraft he was directing. A few minutes later, there was a lull in the radio traffic or as they say, "A pregnant pause." The controller came back on the radio, and talked to the light aircraft, in a soft voice, "Sprecken ze deutsch?" The light aircraft pilot said, "Yahh." The controller then proceeded to absolutely crucify the German light aircraft pilot with the most arrogant tone in his voice, in German, for a full five minutes while we were just rolling with laughter in the cockpit. It was just hilarious!

The deployment from Ohio to Rhein-main AB routed through either Lajes Air Base in The Azores or Goose Bay Air Base in Canada, for a refueling stop. Either way, it was a way to exercise and hone our navigation skills overwater, beyond land based navigation aids. Frequently, this was the only overwater flying we got all year, and we could get our required annual overwater proficiency check during this mission.

Golf at St. Andrews, Scotland
In September 1972, I was on a two-week Creek Party tour. Labor Day fell in the middle weekend. After the first week, all of our aircraft were moved from Rhein-main AB, Germany to Mildenhall AB in England. The Air Force had an exercise going on, and we were in the way. Our navigators had planned a car trip from Mildenhall AB to St. Andrews in Scotland to play golf at their famous golf

course, for the middle weekend. As aircrews finished their required flights in the first week, they flew their aircraft to Mildenhall. My crew had the last flight of the week, a late Friday night refueling, and didn't rotate to Mildenhall until Saturday morning. The rest of the navigators were ticked because they had to wait for me on Saturday. It was a long drive to St. Andrews, with an overnight stay along the way. When we arrived, the other navigators got in their golf. While I didn't play golf, a peer and I walked the town and had a good time.

Screwed Up Rendezvous over the English Channel

On Labor Day, we were tasked to participate in a two-cell Mass Gas (Two waves of tankers for a formation refueling of two waves of NATO F4 Phantom aircraft) over the English Channel towards Norway. The receiver aircraft were on a long deployment from Germany to Bodo, Norway, in the Arctic Circle. My crew was in the second wave of tankers, and took off 45 minutes after the first.

We self-navigated over the English Channel to the orbit point, and established radio contact with the second wave of F4 fighters. As the first wave of fighters approached, the fighter leader declared over the UHF (Ultra High Frequency) radio, "Tallyho," which meant he had the tankers in sight, and pressed on for the rendezvous. The problem was the <u>first</u> wave fighter lead had the <u>second</u> wave of tankers in sight, at the Orbit, not the first wave of tankers that had departed the Orbit for the Air Refueling Control Point (ARCP). Everything deteriorated from that point onward…all screwed up, as they say, but somehow in the end it got straightened out.

Engineer, Set Refueling Power

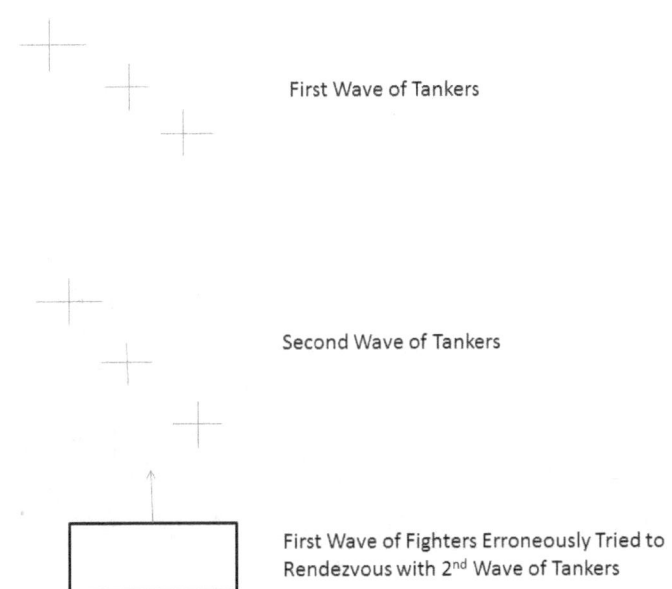

First Wave of Tankers

Second Wave of Tankers

First Wave of Fighters Erroneously Tried to Rendezvous with 2nd Wave of Tankers

Photo by Author

Our aircraft was echeloned off to the right of the lead tanker by 45 degrees. Lead was hooked up to its F4 Phantom fighter over the English Channel.

Photo by Author
F4 Phantom being refueled over Germany (Operation Creek Party)

Rhein-main AB

Rhein-main Air Base, Germany

Engineer, Set Refueling Power

Ohio Air National Guard KC-97L on display at Wright Patterson AFB Museum

Critical Fuel Situation

Six aircrews and aircraft were scheduled to fly to Rhein-main AB for a Creek Party tour. Five of the crews were made up of people who were either self-employed or government workers; they departed on Thursday, November 19, 1971. The next day the remaining crew consisting of people who worked for a living outside the government, including me as navigator, was assembled to depart after work.

The pilot was an outstanding instructor pilot who was an executive at Proctor & Gamble in civilian life. The copilot was a sub-par pilot who had to be matched up with an experienced pilot, in order to stay out of trouble.

When we reported for flight planning, we were told one of the other aircraft lost use of an inverter on the previous day's deployment, and made a precautionary landing at Lajes AB in The Azores. (An inverter is an electrical component that converts DC to AC power.) We were ordered to fly to Dover Air Force Base, Delaware, stay overnight in crew rest, pick up the replacement inverter and deliver it to Lajes AB on Saturday. We were then to refuel and fly on to Rhein-main AB in Germany. The flight to Dover Friday night, after work, was uneventful.

Saturday morning we checked with the weather station at Dover AFB, and were told Lajes AB was closed, due to high wind conditions exceeding the cross-wind capability of our aircraft. Also, the control tower was abandoned because the personnel were concerned the windows would blow in. We "crew rested" all day with occasional calls to the weather station to see if there was any improvement in the weather at Lajes.

Finally, late afternoon, the weather station informed us at our arrival time, assuming a launch within an hour, the winds were forecasted to have died down sufficiently to be able to land. On this basis, we filed our flight plan, fired up all six engines, and took off at 2040L (8:40 P.M.) Forty minutes later, over Cape Cod, my airborne radar failed. As we were briefed about occasional thunderstorms along our route over water, I called the pilot on interphone and stated I needed the radar to circumnavigate the cumulus buildups. We dumped 10,000 lbs. of fuel over water to get down to landing weight, and diverted back to Dover AFB to fix the radar.

At 0215, local time, on November 22, 1971, with the radar fixed, we were again on our way. Now the reader must understand the Air National Guard directed the KC-97L use all six engines in flight, all the time; four piston engines and two jets. The Pratt & Whitney R4360 piston engines burned approximately 850 lbs. of fuel per hour, each, at cruise power, and the GE J-47 jets burned approximately 1500 lbs. of fuel per hour, each. That totals up to 6400 lbs. of fuel consumption per hour, which meant the KC-97L was a gas hog.

Whenever we transitioned a Military Airlift Command (MAC) base, I requested a computer flight plan be generated the night before departure. If nobody dropped the ball (which happened frequently), the flight plan was waiting for us when we reported for flight planning the day of the flight. This flight plan was tailored for specific MAC airlift aircraft. Since MAC didn't "own" any KC-97L aircraft, I requested a computer flight plan for the MAC C-133A turboprop airlift aircraft. Its flight profile was sufficiently similar to ours, with the exception the C-133A needed to "step climb." It was too heavy to climb out to its cruise altitude until some fuel was burned off at lower altitudes. This didn't affect my plans.

"Lead" Tanker with No.2
Echeloned to right 45 degrees

Lead Tanker Refueling F4

F4 Phantom taking on fuel

KC-97L Nav Station

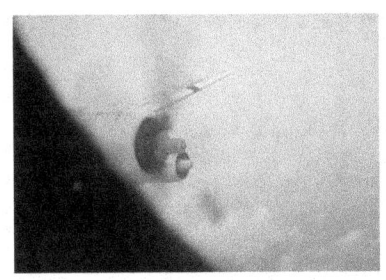

Number 1 piston engine and J1

C-133 aircraft
KC-97L flight profile was similar

"Creek Party" out of Rheinmain AB, Germany

I never understood why the National Guard required all six engines be used, continuously. When I crewed on the KC-97G in SAC, we had only the four piston engines. If we had shut down the jets, we would have had more fuel reserves.

Diverted to Santa Maria Island

We took off again and climbed to 25,000 ft. The flight engineer set cruise power for 275 KTAS (Knots, True Airspeed). There was no overcast, so I used night celestial three-star fixes, Loran, and pressure pattern for navigation all the way across the ocean, until I picked up the land mass of The Azores within 100 nautical miles, using airborne radar.

The flight over water was uneventful, until we came within radio range of Lajes AB. We were shocked to learn the winds had not died down as forecasted, and the airfield was still closed to incoming traffic! The pilot asked me to identify the alternate field and time enroute. I responded it was Santa Maria Island and the time was 32 minutes. The copilot contacted Santa Maria on our UHF radio to request permission to land, but permission was denied. I reported our second alternate was Lisbon, Portugal (841 nautical miles; 2:37 hour's enroute time). Without being asked, the flight engineer quickly said, "No Sir" to the remaining fuel on board. We just burned too much fuel with the jets operating, flying over the ocean.

Engineer, Set Refueling Power

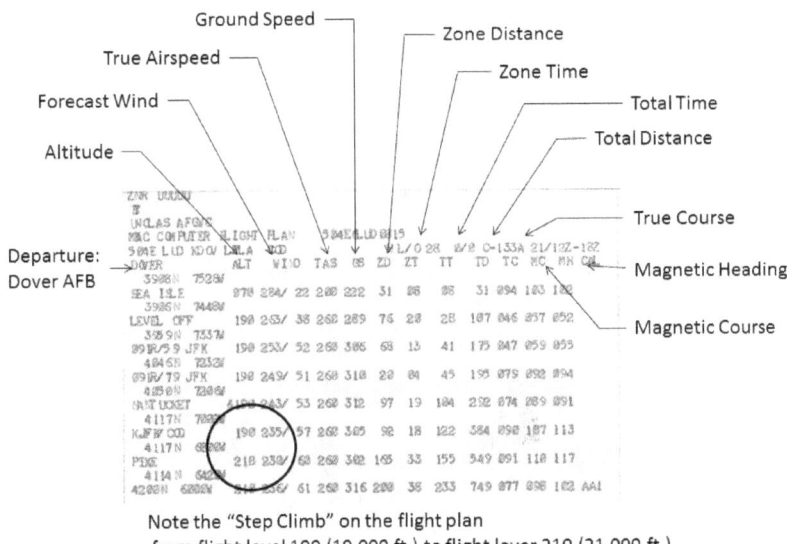

Note the "Step Climb" on the flight plan
from flight level 190 (19,000 ft.) to flight lever 210 (21,000 ft.)

Computer Flight Plan for C-133A Aircraft

Douglas C-133A aircraft
Ref: nationalmuseum.af.mil

```
ZNR UUUUU
B
UNCLAS AFGWC
MAC COMPUTER FLIGHT PLAN    504E4LD0015
504E LLD KDOV LPLA  COD                 H L/0 28  B/0 C-133A 21/12Z-18Z
DOVER           ALT  WIND  TAS  GS  ZD  ZT    TT   TD  TC  MC  MH CAL
  3988N  7528W
SEA ISLE        878 284/ 22 200 222  31  08   08   31 094 183 182
  3986N  7448W
LEVEL OFF       190 265/ 38 260 289  76  20   28  107 846 857 852
  3959N  7337W
091R/59 JFK     190 253/ 52 260 305  68  13   41  175 847 859 055
  4046N  7232W
091R/79 JFK     190 249/ 51 260 310  20  84   45  195 879 092 894
  4050N  7206W
NANTUCKET      &190 243/ 53 260 312  97  19  184  292 874 889 091
  4117N  7002W
KJFW COD        190 255/ 57 260 305  92  18  122  384 890 107 113
  4117N  6800W
PIKE            210 230/ 60 260 302 165  33  155  549 091 110 117
  4114N  6420W
  4208N  6000W  210 236/ 61 260 316 200  38  233  749 077 038 102 AA1
  4200N  5000W  230 283/ 52 260 310 446 127  400 1195 090 114 112 AA2
KJFK/LPAZ       230 299/ 41 260 298 453 131  531 1648 858 121 118 AA2
  4100N  4200W
  4000N  3500W  230 269/ 52 260 310 236  46  617 1884 105 126 129 AA3
LIMA            230 255/ 67 260 318 196  29  646 2040 101 121 126
  3950N  314 W
RGRESRB         230 253/ 72 260 317  24  85  671 2064 107 126 133
  3923N  311 W
GRACIOSA RBN    230 251/ 73 260 324 148  27  718 2212 097 116 122
  3905N  280 W
LAJES           230 251/ 74 260 309  48  29  727 2260 115 132 141
  3845N  2709W
     FWF  848 WF1   850 WF2   847 ETP 306   PG 011 TOGW 230    85
 A1 SANTA MARIA      253/ 76 260 280 148 8032 LPAZ
  1-0727 053108  2-0037 003692  3-2804 056792  4-0032 5-0115 6-0015
  7-1006 068555  9-200020  10-001580  11-070055  B/0-054600
 A2 LISBON/PORTELA  261/ 61 260 320 841 0237 LPPT
  1-0727 053108  2-0244 024378  3-0311 057478  4-0237 5-0115 5-0015
  7-1218 085 51  9-200000  10-001580  11-062061  B/0-054600
 A3 RGTA             261/ 53 260 310 1022 0316 LERT
  1-0727 053108  2-0048 004770  3-0815 057870  4-0316 5-0115 6-0015
  7-1301 064338  9-200200  10-001500  11-085838  B/0-054600
B
12 52
```

Computer Flight Plan, Dover AFB, Delaware to Lajes AB, Azores; Time in Route was seven hours and 27 minutes The computer flight plan that we requested was generated for the Military Airlift Command (MAC) C-133A Aircraft. Note that the plan was for the aircraft to "Step Climb." The aircraft initial cruise was at FL190 (19,000 feet altitude) with the altimeter setting at 29.92. After 1 hour and 22 minutes, the aircraft planned a climb to FL210. After 2 hours and 33 minutes, total time, it would climb to FL230. This "Step Climb" was necessary for the C-133A because of the aircraft weight/engine power relationship. The aircraft was simply too heavy to climb to FL230 right after takeoff.

Now there was a quick huddle in the cockpit with an instructor pilot who was dead-heading and the rest of the crew. The aircraft commander made the decision to declare a fuel emergency, and informed Santa Maria we were landing. By declaring a fuel emergency, Santa Maria could not refuse our request for landing permission.

After we landed and parked the aircraft, I made a note the indicated airspeed instrument read 60 knots. We were met at the aircraft by a Portuguese baggy-panted guard with a rifle who was not particularly pleased to see us. He drove us to a hotel to spend the night. I remember sitting in the lobby with the crew and commenting on the 20-watt light bulbs; it was difficult to see. We learned later the Portuguese government allowed Soviet Union ships to visit Santa Maria Island, and the government did not want any problems with an aircraft from the U.S. military.

A Case of Wine
While we were in the hotel lounge, another group of Air Force personnel came in. They had the same problem with the weather. Their C-54 aircraft was enroute to Lajes AB and had to be diverted into Santa Maria Island, just as we were, after a shopping trip to Europe for a general's wife.

The general's aide thought he would impress us with his knowledge of the local wines. I always purchased a case of Rose Mateus bubbly red wine to bring home, whenever my crew stopped in Lajes. This general's aide told us the locals laughed at the Americans, as they considered Rose Mateus as the *dregs*. Real men drank Terras Altus wine. I stored that information in my mind, and later bought the allowed limit of three bottles of Terras Altus. When I returned home and tasted Terras Altus, I threw it all out, as it was awful. So much for the bravado of a know it all.

KC-97L tail number 918 had just diverted to Santa Maria Island in the Azores island chain. Author is on the right. Photo by Author

The next day, with the winds reduced, we were transported back to our aircraft by the same baggy-panted local. As we climbed the entry ladder to our aircraft, the guard spotted the two large coffee jugs in the cockpit galley. By now, the coffee was at least two days old, and stone cold. The guard motioned to the coffee, and we motioned him to go ahead. He drank a cup of what must have been terrible coffee, and made an approving gesture by smacking his lips. I groaned as I thought about how awful the coffee must have tasted. The trip to Rhein-main AB in Germany was uneventful.

I returned from these deployments with a sense of professional fulfilment. This flight was probably my only over-water flight for the year. Most of our other flights were over land, where it was difficult to screw up the navigation.

Engineer, Set Refueling Power

181st Air Refueling Squadron; KC-97L S/N 53-0360
F4 Phantom Refueling off an Air National Guard KC-97L Tanker
Ref: commons.wikimedia.org

KC-97L from 125th Air Refueling Squadron, Ohio Air National Guard
Note blue smoke from engine exhaust due to engines running in "auto-rich" setting

Engineer, Set Refueling Power

KC-97L

145th Air Refueling Squadron Patch, the Tasmanian devil, on left; Tactical Air Command Patch on Right, the Gaining Command on Call up

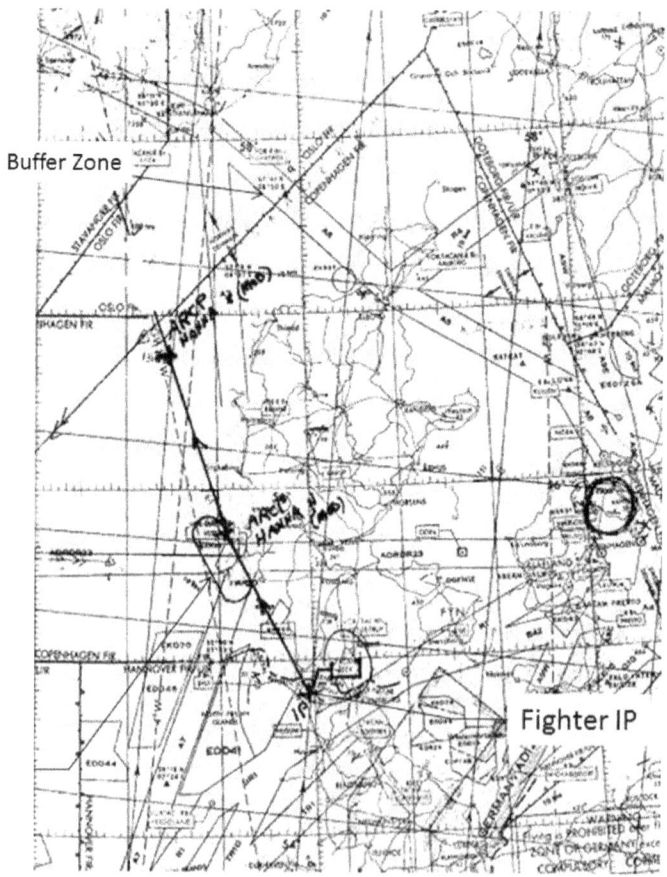

Hannah Refueling Track, West Germany
Chart provided by Author

Cessna A-37 Light Attack Aircraft

North American F-100 Super Sabre Fighter Aircraft

Brutally Cold "Drill" Weekend at Lockbourne AFB, Columbus, Ohio

It was the weekend for our "Drill" and the forecast was for brutally cold weather. I had driven up Friday evening and checked into the BOQ. Saturday morning I did not expect my six-volt battery powered 1967 Volkswagen to start. It was so cold when I pushed down on the clutch pedal, it felt I pushed molasses in the hydraulic clutch. What the heck, I had nothing to lose.

I turned the ignition key to the start position, and the engine made a sound like, "Ruh.....Ruh.....Ruh" followed by one of the four cylinders firing, "Chugg......Chugg.......Chugg.....Chugg". Soon the second cylinder also fired, "Chugg Chugg.....Chugg Chugg......Chugg Chugg". Subsequently both the third and then the fourth cylinder fired and the engine started to warm up. My car was the only one in the parking lot to start that early morning. I couldn't believe it.

Engineer, Set Refueling Power

Cross Country to Bermuda

During one training weekend (drill), just before leaving to come home, the air technician navigator told me I was due for my annual over-water proficiency navigation check, and there were three aircraft scheduled to fly to Bermuda the following Friday through Sunday. I was to be on one. I didn't complain, as I really wanted to go. I told both my wife and my boss I was required to go. I had a good navigation leg to Bermuda, and a great weekend. One of the other two navigators picked up two school teachers in the officer's club bar, and fixed up the navigator from the third aircraft. The next morning, the other navigator had a loud running beef. Apparently, they both got lucky Saturday evening, but the other nav must have had to perform all night long, from what I could glean from the loud conversation, and he was not too happy about it.

On Sunday morning, May 23, 1971, when we prepared to come home, the number three engine (right side, inboard engine) on our aircraft wouldn't crank. The starter had failed. Well, there was an option; we could pile on one of the other two aircraft and come home, but it would have required an aircraft returning with two crews and maintenance people, to ferry the formerly *sick* aircraft home after repair.

We decided to try a windmill start. This encompassed starting the other three piston engines, doing an engine run-up and checkout, followed by starting the two jets. We made a high speed run down the runway to get the *dead* engine windmilling. We did and it did. The flight engineer was able to catch the engine while windmilling and get it running (somewhat like jumping the clutch on a car with a dead battery). Once the engine started, the pilot stopped the aircraft on the remaining runway, and returned for checkout and warm-up of the *offending* engine.

We flew from Bermuda Naval Air Station to Charleston AFB to declare customs. Upon arrival one pilot sat in the "seat" with the number three engine running constantly, while the rest of the crew went into base operations to declare customs,. We <u>were</u> going to get home, because we all had to go to work on Monday. We did not chance shutting down number three engine!

Worst Rendezvous of My Career

We took-off on a typical Friday evening to refuel North American F100 Super Sabres, from the Mansfield Airport (Ohio Air National Guard). When we arrived at the Orbit, we found out the fighters had taken off early and were circling the Orbit waiting for the refueling. Now, this was not the usual procedure for conducting a rendezvous, and to be frank, I was not sure how to get the fighters behind us.

The next thing I knew, the fighters were on a head-on collision course with us. The pilot requested the fighter's altitude, and it was 50 feet higher than us. I could virtually see the hairs on the back of the pilot's neck standing up, as he and the copilot tensed up and strained to see the oncoming fighters. Zoom, they went right over the top of us at a closing speed of approximately 600 Knots! By the time I got them turned around and heading in the same direction, the fighters were way behind us, and in a tail chase. It was a screwed up mess. This was the worst rendezvous I had ever been involved with.

Most Rewarding

My time in the Ohio Air National Guard was the most rewarding of my flying career. My "job" at GE Large Engine Division was a bust, and the flying in the KC-97L saved my sanity during the few years I spent at the Evendale GE Plant.

Putting the reader in the cockpit

The following was written to give the reader a feeling about what it was like on a deployment to Germany. The crew performed the "Preflight Inspection" on the aircraft, started engines 30 minutes before takeoff, on my time hack, and ran up the engines for checkout. While counting down the time for takeoff, my armpits got a little sweaty, and I had to admit I was nervous. There was nothing like the responsibility of being the only one on the crew who could keep the aircraft on course and make the coast-in point.

Engineer, Set Refueling Power

I counted-down the time to takeoff. At the one minute warning to takeoff, the pilot taxied into position on the active runway and set the brakes. I gave the 30 second call and the pilot directed the flight engineer to set "barometric" engine manifold pressure (power). At my 10 second call, the pilot announced, "Pilot's Throttles," and set takeoff power; the copilot toggled the spring loaded switches controlling the two jets to 100% power. Now, the aircraft just strained at the "leash." There was a powerful "throb" throughout the aircraft. This was the most exciting phase for me. "Five, four, three, two, one, hack."

The pilot released the brakes, and the aircraft started its acceleration down the runway. Everything on my navigator's desk slid aft, and I had to trap all the "stuff" before it fell to the floor. As the aircraft approached 60 knots on the runway, the copilot announced, "60 knots" and right at 60 knots, "Now." I started my stop watch and, as the pre-announced time approached, stated over interphone, "Time," and "Hack" at the actual time. On this flight, at our heavy weight, the acceleration check was 17 seconds. This was from 60 to 100 knots airspeed! Even though we had the jets, the aircraft was fully loaded with maintenance personnel and spare parts

After several hours over water, using celestial navigation (three-star fixes) aided by pressure pattern, Loran and Consolan, I adjusted the radar controls to pick-up the coast-in land mass; the radar range was set to 100 miles, the antenna sweep was adjusted to sector scan from full scan, the antenna tilt was adjusted up and down until the coastline of the land mass was highlighted. I then informed the pilot Lajes AB was on the radar scope, and he turned-on his repeater radar scope.

This experience contrasted with my unrewarding job at GE, Evendale, as I was a small fish in an ocean of talented technical people. My job had overlapping responsibilities with other engineers, and I felt if I died, the job would get done by others, and my absence would not even be noticed. I just didn't have a good feeling about my contributions. The Air National Guard provided the job satisfaction I lacked at GE. The long overwater navigation legs to and from Europe relied on my professional skill, and I felt great upon arrival.

Possibility of Layoffs at GE

I started work at GE Evendale with high hopes of working on jet engines. It didn't take long to realize GE was a bloated bureaucratic organization run by a cult of personalities. Perhaps every large organization was similar, but it was a shock to me. I quickly became depressed.

Within the first year of employment, the aerospace recession of 1970 started. Orders for commercial aircraft and their jet engines plummeted, and GE went into full cutback mode. The GE Product Support Operation, set up to support the new CF6 engine (in the new DC-10 aircraft) when it went into production, was decimated. The entire basement in Building 800, where this organization was located, was emptied and the lights turned out.

I turned to the OANG to keep my spirits up. I figured if I was laid off, I could fly as much as possible to maintain some semblance of a reasonable income. My salary at GE at the time was $12,000 per year (1970), and this was comparable to what I would earn if I was participating in the military full time (which I was not allowed to do, but I could come close).

Back-to-Back Additional Flying Training Periods (AFTP's) in the OANG

Flight crew personnel were authorized to fly 32 AFTPs per year. Most flying training was scheduled on Friday nights and Saturday Mornings. The rule to accomplish an AFTP, and be paid, was to perform a duty day of four hours, minimum, and fly an average of 2.5 hours during that duty day, over a six month "quarter." I signed up to fly most weekends. Generally, I left work somewhat early on a Friday afternoon, drove the 100 miles to Columbus and checked-in to the Bachelor Officer's Quarters (BOQ) to get a room for the night.

In those days, each room housed two officers, so I didn't know who I was bunking with. I believe I paid $1.00/night for the room. Next, I drove to the squadron operations building to accomplish the flight plan. There, I checked the status board to see who I was flying with, and details on the mission. An example is as follows:

Date: September 17, 1974

Crew: (P) Collins; (CP) Brown; (Nav) Dornfeld, (FE) Link; (BO) Devers

Mission: T/O (Takeoff Time) 1800L; A/R (Air Refueling Area) Tea Shop, ARCT (Air Refueling Control Time) 1930L

Receivers: (2) F-100's from Mansfield Airport; Tactical Call Sign, Keeper 23 and 24

Fuel Offload: 5000 pounds of JP4 each

Navigation Training: Night Celestial

The pilot and navigator drove to base operations where the weather station was located, to get a briefing on the forecast weather enroute, and the winds aloft. I then completed my flight plan with these forecast winds to calculate compass headings, times between points and total times.

The Notice to Airmen (NOTAMS) were reviewed. These were posted on a board, and consisted of things like a certain radio facility was inoperative, or there were to be fireworks in a certain area. I obtained a time hack from the U.S. National Bureau of Standards Radio Station, WWV, which gave a precise time hack based on the atomic clock, every five minutes. Even better was the Canadian Station, CHU, which gave a time hack every minute. It was necessary to have accurate time, because celestial navigation was based on the location of all celestial bodies at a particular time. An error of four seconds could result in a navigation error of one nautical mile.

The copilot picked-up the flight lunches and the crew reassembled in the squadron operations building for a mission briefing. This was a formal briefing with all aspects of the mission discussed. I briefed the route and the rendezvous/ refueling data.

We picked up our "Personal Equipment (PE)," from the PE shop, which consisted of our oxygen mask and parachute harness. The parachutes were the chest pack type, and were already aboard the aircraft. In a bailout situation, we were to attach the chest pack parachutes to the harness as we were about to jump out of the aircraft.

The crew met outside where we were driven to the assigned aircraft in a "bread" truck. At the assigned aircraft, the pilot reviewed the AF Form 781, which was the aircraft maintenance form. Any discrepancies discovered on the previous mission were written and the disposition by the maintenance crew was noted. The pilot then released the crew to accomplish its preflight inspection.

Typical Bread Truck
Vehicles similar to these were used to transport aircrews to their aircraft

The "Start Engine" time was 30 minutes before takeoff. Everything was precise. I announced over interphone "30 minutes to Engine Start." Then "10 minutes, five minutes, 30 seconds, and five, four, three, two, one, and hack." On my hack, the flight engineer pushed the starter button on number three engine. The sequence for engine start was number three, four, two and one. Engines were numbered from left to right, as viewed from the pilot's seat. Number three engine was the right inboard engine as viewed from the rear. If there were more than one aircraft for this mission, and you were standing where you could see all the aircraft on the ramp, you would see number three propeller on all the aircraft on the mission start to turn precisely at the start engines time.

After engine start (piston engines), once oil temperatures reached minimum, the pilot called for the "Taxi" checklist. He rang the alarm bell, which was the means for anticipating a crash landing or bailout, to test its operation. I announced on interphone, "Alarm bell loud and clear, (landing gear) down locks and boom pin stowed, lower forward entry door closed and locked, Nav ready to taxi."

The pilot taxied the aircraft to the run-up pad and, with the flight engineer, performed a thorough engine and propeller checkout. I continued to give the pilot a heads up on the minutes remaining to takeoff time. The copilot started the two booster jets, designated as J1 and J2. When the pilot announced the "Before Takeoff Checklist," the flight engineer briefed the acceleration check time. At my announcement of 30 seconds to takeoff, the pilot announced, "Pilot's Throttles."

Once our takeoff clearance was received from the tower, the pilot taxied onto the active runway and set the brakes. At my 15 second call, maximum power was set on the piston engines and the two jets were toggled to 100% power. The aircraft strained against the brakes. There was a vibration "throbbing" throughout the aircraft. "Five, four, three, two, one, hack" and the pilot released the brakes. The aircraft lunged forward with the power of four piston engines (3500 shaft horsepower, each) and two jets (about 5000 lbs. of thrust, each). Takeoffs were most exhilarating.

A typical acceleration check was five seconds. This was the maximum time it took to accelerate the aircraft from 60 to 100 knots airspeed, while on the takeoff roll. As the aircraft approached 60 knots, the copilot announced on interphone, "60 knots." At exactly 60 knots, he announced "now." I started my stopwatch. As five seconds approached, I said, "Time." At five seconds, exactly, I said, "hack." The pilot and copilot compared their airspeed instruments to confirm the speed was at least 100 knots. Each had their own independent airspeed instrumentation. I never had a mission abort due to a failed acceleration check. I have to admit in SAC, on active duty, the aircraft, without jets and heavily loaded, had acceleration checks of as much as 17 seconds.

The mission itself was routine. Unlike when I was on active duty in SAC, the aircraft and engines were reliable. I cannot recall having to shut down an engine while I was in the Ohio Air National Guard. On active duty, engine shutdowns were frequent, due to both the experience factor (constant turnover of maintenance personnel) and the taxing heavyweight missions we flew.

I navigated to the air refueling area using Airways (radio navigation on "highways" in the sky), conducted the rendezvous with the fighter aircraft, and navigated within the refueling area, using my airborne radar. Once the refueling was completed, and the aircraft engine power was returned to "Cruise" configuration, I accomplished a two-hour night celestial navigation leg back to Lockbourne AFB. Upon arrival and descent, the pilots made a few precision and non-precision approaches to the runway. Sometimes an approach was made with a missed approach and go-around, and sometimes a touch and go was made on the runway. This was followed by a full stop with propeller reverse thrust and mission termination.

My time in the Ohio Air National Guard was the most rewarding of my flying career. My "job" at GE Large Engine Division was a bust, and the flying in the KC-97L saved my sanity during the few years I spent at the Evendale GE Plant.

After the final landing, I "crashed" in the BOQ. In those days, there were two to a room, so I never knew who I was rooming with. Once, after landing and filling out the paperwork to get paid, I returned to the BOQ and unlocked the door to my room. There was a newly commissioned 2nd lieutenant in his bed. He was shocked when I entered, jumped up and attempted to salute. I quickly stopped that nonsense and put him at ease.

The next morning, I did it all over again, only performed a day celestial navigation leg. As I drove home on Saturday about 2:30 P.M., I drove on air as I was so pumped up with the feeling of accomplishment. It gave me what I did not get working at GE. The accomplishments for the two days were:

Two Aircraft Rendezvous

Two Air Refuelings

One Night Celestial Navigation Leg

One day Celestial Navigation Leg

Two days Base and Flight Pay (about $35 net for each day, in 1972)

The air traffic in Europe in this timeframe was congested. The AN/APX-29A rendezvous radar picked up only IFF/SIF Mode 2 signals (the channel for military aircraft, only), and in the United States, only the intended fighter aircraft signal was usually picked up, due to the scarcity of military aircraft in the area.

Complaints about Noise from Rhein-main AB
A small town off the active runway complained about the aircraft noise from both Frankfurt International and Rhein-main AB (They shared the same runways.) The town government put up an audiometer to measure the sound level from each aircraft as it climbed above it, and snapped a photograph of the tail number when the noise level exceeded a threshold. Then the town submitted a bill to the U.S. Government or airline for violating the ordinance.

We were briefed on this and took precautions to avoid being discrepant. We made our takeoff run down the runway and held the aircraft on the runway, building up airspeed. We broke ground way above minimum climb speed, and did a maximum performance climb. What with four reciprocating engines and two jets, we were never caught exceeding the limit by the time we crossed over the town. And the takeoffs were exhilarating.

Cross Country to Bermuda

My experiences with the 145th Air Refueling Squadron of the Ohio Air National Guard in their KC-97L aircraft were the best in my Air Force career. The aircraft were the most reliable, the aircraft performance was exceptional, and the people were great.

I wanted to have my wife, Marilyn, join me in Germany, but my kids were small, and there was nobody to watch them. I committed to her the next Creek Party I went on, I would somehow have her join me. Alas, I got my transfer from the GE Large Jet Engine Plant in Cincinnati, Ohio to the Small Jet Engine Plant in Lynn, Massachusetts, my hometown, and the last trip to Germany didn't come off.

I transferred to GE, Lynn, Massachusetts

I finally secured my transfer to GE, Lynn, Massachusetts. I made an appointment with the wing commander and told him of my transfer. There were no openings in any Air Force Reserve or Air National Guard unit anywhere near my work. I was frustrated. He went out on a limb for me. He called the commander of the New Hampshire Air National Guard unit at Pease AFB, Portsmouth, New Hampshire and made an agreement to allow me to train with this unit for three months, while I was still attached to the Ohio unit. Ohio paid me, and I demonstrated interest and participation, in order to try to worm my way into the New Hampshire unit.

After three months, I met with the commander of the New Hampshire unit. He told me he wanted me, but his hands were tied. He previously signed up two individuals who he sponsored through Undergraduate Navigator Training. These two navigators had a six year service commitment they could not get out of, and the unit had absolutely no openings.

DORNFELD, GERALD R.
017-30-3472FG

CAPT
24 SEP 72

Author's Official Air Force Photo
Ohio Air National Guard

The interesting fact was these two young navigators refused to participate fully in the ANG. Every AFRES and ANG crew member was authorized 32 Additional Flying Training Periods (AFTPs) to maintain flight proficiency. In the Reserve, these days were available, but in the Guard, using these days was mandatory. Additionally, these two individuals refused to pay their annual Air National Guard dues. There was nothing the unit could do to these two navigators.

After I left the Ohio unit, when I transferred to GE, Lynn, the Air

Engineer, Set Refueling Power

Force started moving KC135A jet tankers into the Air National Guard and Air Force Reserve. My old unit, the 160th Air Refueling Group, was among the first to be integrated with the SAC KC-135 tankers.

KC-135 moved into 145th Air Refueling Squadron, Ohio Air National Guard
Ref: en.wikipedia.org

The New Hampshire Air National Guard was equipped with the C-130A aircraft
Ref: commons.wikimedia.org

Engineer, Set Refueling Power

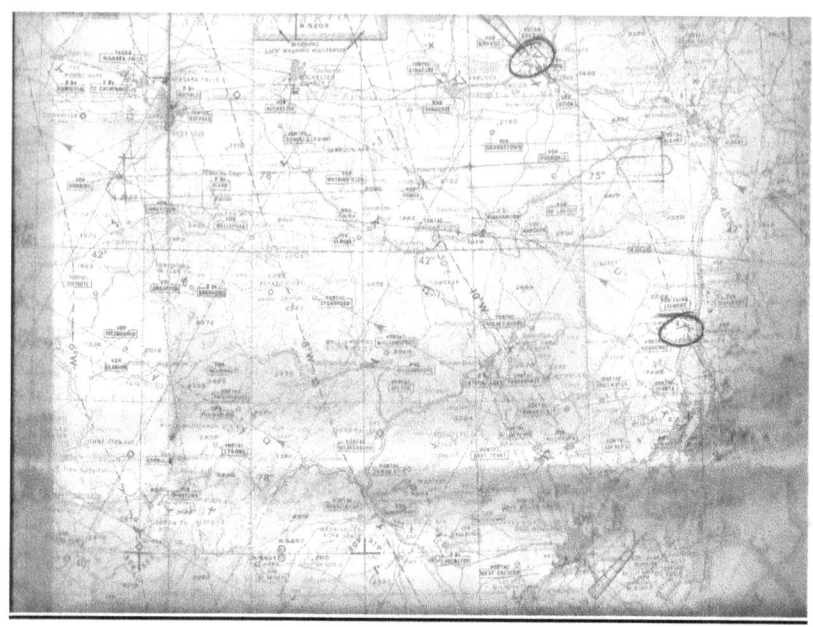

Air Refueling Track across upstate New York

A37 Internet Photo
Ref; http://en.wikipedia.org/wiki/File:OA-37B-1.jpg

CHAPTER 10; CAREER AT GE AIRCRAFT ENGINES

I was still working for Hamilton Standard and was seriously, but unsuccessfully, looking for a career change. My sister worked as a secretary for the GE Lynn, Massachusetts Thomson Lab, which was the metallurgy laboratory for the facility. GE Lynn was the engineering and manufacturing division producing small aircraft engines. Her boss took my resume and circulated it, both in the Lynn and the Evendale (Cincinnati, Ohio) divisions. I was asked to come to Evendale for two interviews; one with the CF6 Commercial Engine Group, and one with the F-101 Military Engine Group. To make a long story short, I was hired by the Evendale, Ohio plant, as a CF6 Engine Installation Engineer, at $12,000 per year, a modest increase in salary. This was how I joined GE.

My Move to Cincinnati was a Nightmare (Nov 1969)
I needed to report to my new job at GE, but my house in Bloomfield, Connecticut had not been sold. I made a decision; Marilyn and the children would stay with the house until it sold, and I would go ahead to Ohio. I left my more modern car with Marilyn, my 1967 Volkswagen Fastback, and drove my second car, the 1963 Volkswagen Bus, to Ohio. This latter car was a rust bucket, but it ran fine. However, the VW bus had no heat, even though it had an auxiliary gas heater, which was inoperative. I had no money to have the gas heater repaired.

The timeframe was November 1969, and it was cold. I dressed in my Air Force winter thermal flight suit underwear and thermal insulated boots. All this insulation was insufficient to keep me warm. I was cold! While driving this used bus to Ohio, the transaxle started to howl as I passed Columbus, Ohio, about 100 miles from Cincinnati. By the time I arrived, I was afraid to drive the bus any further. I sold the car, cheap, to a youngster, who really wanted it.

The GE relocation allowance for a new hire was not as good as it was for a transferred employee. GE paid for a motel for two weeks before I was on my own, and maintaining two homes was too much for me financially. I decided to move my family to an apartment in Cincinnati. As we settled into our apartment, the telephone was just activated and it rang; it was the realtor from Bloomfield, Connecticut, with an offer of $30,000 from a buyer. From a financial viewpoint, this came just in the nick of time, and I grabbed it!

I joined GE in the middle of the famous strike of 1969. I was nervous about crossing the picket line in order to join the company, but was able to do this without a problem. The strike lasted for months, and the daily report to work became an unpleasant adventure. Some days were easy, and others were a nightmare. Some days I had to turn around and go home and try to enter the plant later in the day. As my boss related to me, I was expected to attempt to come to work; if I was turned away by the pickets, I was expected to try again later in the morning!

I was thrilled to be going to work for GE, as I foolishly thought I simply was going to move into jet engines, doing the same work as I had been doing at Hamilton Standard, because the job titles were so similar. Alas, this was not to be. At Hamilton Standard, I had visibility to upper management in my work interfacing with customers. However, at GE, there was an entirely different organization of higher-level engineers, called the "Project," doing the customer coordination. My role was to take direction from the Project, and coordinate customer inquiries within the engineering organization. Basically, I pushed paper, and I was depressed.

My work involved coordinating the CF6 engine fuel system and condition monitoring interfaces with Douglas (through our Project organization) on the DC-10-10 aircraft installation. I also coordinated the aircraft flight deck instrumentation range and accuracy for both the CF6-6 and CF6-50 engines. In addition, I implemented the CF6 Installation Approval Plan, within engineering, by coordinating the requirements for analyses, component test and flight test data with Douglas (again, through our Project organization).

Engineer, Set Refueling Power

A Jet Engine Primer

Just as I provided Isaac Newton's Second Law of Motion, while talking about propellers, the same relationship can be made for a jet engine. Whereas a propeller develops thrust by accelerating a large amount of air by a small amount, a turbojet engine moves a small amount of air by a large amount.

The same equation holds; $F=MA$. In the case of a more modern turbofan engine, there is a compromise in the airflows; the fan produces most of the thrust by moving a larger mass of air a large amount, but there is a residual thrust from the core engine developed by moving a small amount of air a relatively large amount.

Bad Experience with State of Ohio Licensing of Professional Engineers

As I wandered around the engineering office area, I noticed many engineers had a "Professional Engineering License" mounted in a picture frame on their cubicle wall. This was my introduction to state licensing for professional engineers. One had to take an initial "Engineering in Training" examination (EIT) administered by the State of Ohio, followed by the full examination on engineering subjects later. I was not required by GE to be professionally licensed in order to work in aerospace, but I thought it would be a good thing to do so. I took the EIT engineering review course after hours, given by GE, to prepare for the EIT test.

I sent my application fee to the State of Ohio, and waited for their response with a date and place to take the exam. This was the time frame in which I had just moved to Ohio, was flying with the Ohio Air National Guard in their KC-97L aircraft, while still assigned to the Air Force Reserve C-124 aircraft unit at Westover AFB, Massachusetts.

The date for the exam was nearly upon me and the administrative agency had not contacted me. I called the State of Ohio office in charge of administering it. They informed me I did not hear from them because I had graduated from an engineering school not approved by the Engineer's Council for Professional Development. I was shocked, and angry. I called Boston University Engineering School, and they confirmed they were not yet accredited, but they were working on it. When it was received, it would be retroactive to all earlier classes. Funny, <u>nobody</u> bothered to tell me this information when I attended the university, but they were happy to take my money!

Much later, I found out Boston University did indeed receive the accreditation from the Engineers Council for Professional Development. By then, however, I lost all interest in studying and taking exams to prove to the State of Ohio I was a competent engineer, long after I proved the same to Boston University School of Engineering!

McDonald Douglas DC-10 Internet Photo
http://en.wikipedia.org/wiki/File:Ariana_Afghan_Airlines_DC-10_Fitzgerald.jpg

Right at this time, my flight status became clarified, with the help of Senator Robert Taft of Ohio. I received all the money owed to me for flying with the Ohio Air National Guard, and I received my commission as a captain in the OANG. The heck with engineering licensing; <u>I was going to fly with the Air National Guard</u>! This was a decision I never regretted, as it was the only shining light in my Ohio experience.

GE CF6 Engine
http://en.wikipedia.org/wiki/File:Turbofan640.jpg

Ground Floor of Affirmative Action Program

During my first year at GE, I was among the employees selected at random to attend a "Round Table" upper management discussion. The subject was a new initiative, called "Affirmative Action." The GE Aircraft Engines Vice President, Gerhard Neumann, made the introduction, followed by his Manager of Personnel. The latter outlined a company program to outreach to the black community and make opportunities otherwise weren't happening. He ended his presentation with the statement I remember to this day, and I quote, "This may seem like reverse discrimination, and it is, and that is the way it is going to be!"

And so began the reverse discrimination against those of us who had nothing to do with problems of the past. Basically, my generation was to suffer to compensate for the errors of past generations! Going forward, I never observed anyone in upper management giving up their positions in the name of Affirmative Action!

CF6 Installation Approval Plan

My primary assignment was to implement the GE Installation Approval Plan, a plan outlining the logical steps (tests and analyses) leading to GE formally approving the installation of the CF6-6 engine in the DC-10-10 aircraft. This approval was a contract requirement with McDonald Douglas, the aircraft manufacturer. Once GE signed off on the approval, the company was at legal risk for fixing deficiencies in the engine design, as well as fixing the impact of deficiencies requiring aircraft modification.

The plan had been roughed out by my unit manager before I was hired. I inherited the plan, and had to flesh it out. My unit manager required me to update the plan monthly, with progress, such as which test had been accomplished, and which analysis had been made, and the conclusions. I had to meet with the managers of all the engineering specialties (Fuel Controls, Lube Systems, Structures, etc.) to get their updates. These managers were so busy and under such pressure, the last thing they wanted to do was meet with me monthly. This made for an untenable situation for me, and I hated it.

Flight Operations Engineering Career (Not)

I met two engineers in the flight operations engineering group. Their responsibilities were to fly with their assigned airline, when the aircraft went into service, in the cockpit jump seat, in the DC-10, and monitor how our engines were used. The first two customers for the DC-10-10 were American Airlines and United Airlines. The flight operations engineers were tasked to come up with suggestions on how to maximize the life of the engine on wing, etc. This was definitely what I wanted to do, and it seemed a good fit with my flying background. I made an appointment with the manager of this group, for an information interview.

He was a former flight engineer with BOAC airlines, in the days when this airline was equipped with aircraft with piston engines and propellers. We had a lot in common, and told each other *war stories.* He told me he was definitely interested in bringing me on board, and I would be on the top of his hiring list, but there was a hiring freeze imposed at the time. The next thing I knew, there was a minority engineer assigned to the group. When I inquired, I was told the manager *had* to take the new hire. The affirmative action plan was being implemented. The thing that stuck in my throat was the minority engineer didn't even want the job!

KC-97L in Ohio ANG Lettering
Ref: http://www.airpowerworld.info/other-military-aircraft/boeing-kc-97-stratotanker.htm

Aerospace Recession of Circa 1970
The big aerospace industry recession occurred about one year after I came to GE, in Evendale, Ohio. The contract GE had for the U.S. commercial supersonic aircraft engine development had been cancelled, earlier. The rationale was the government-mandated aircraft noise limits could not be met with the afterburning turbojet engine GE was developing. It left the field open for the British and French Concorde with the Rolls Royce engine. The Concorde couldn't meet the noise requirements either, but the British and French got around the new regulation by not using afterburner over land, only over open-ocean.

Next, many of the advanced orders for the DC-10-10 and DC-10-30 were cancelled by the airlines. On top of all this bad news, the GE F-101 engine was not selected to power the new Lockheed F-16 single engine fighter; the award went to Pratt & Whitney with their F-100 engine. If that wasn't enough, the advanced bomber program, the B-70, was cancelled. This bomber was equipped with the GE J-93 afterburning jet engine. GE went into full cutback and layoff mode.

Here I was, a relatively new hire engineer, with one year of service, a car payment on my new 1970 Chevrolet 4-door hardtop, and a house under construction in Forest Park, a family consisting of a wife and three children to support, and nobody to look out for me! Engineers were dropping like flies all around me. Every Friday afternoon, if your boss asked to see you, you knew it was the *kiss*. Several engineers in my operation were selected for layoff. To this day, I do not know what saved me from the layoffs; maybe it was because GE had so recently paid to relocate me to Cincinnati.

I did witness excellent engineers scrambling when they got the notice; they called the personnel department in other GE divisions, in an attempt to find work. As soon as personnel heard a GE Evendale engineer was calling, they were suddenly unavailable. Life was cruel, and I was a nervous wreck! My fallback plan, if laid off, was to volunteer for as many short active duty tours with the Ohio Air National Guard as possible, and perform all my 32 authorized AFTPs. My pay with the ANG was roughly comparable to my GE salary, if I were paid full time with the Guard.

GE Aircraft Engines Vice-President Gerhard Neumann
It would be good to make reference to the GE Aircraft Engines Vice President at the time, Gerhard Neumann. He was an interesting person. He apprenticed as a machinist in Germany before the WWII, before getting his engineering degree, as was the practice back then, in Germany. Gerhard volunteered to join the American Volunteer Group in China, the "Flying Tigers," before the Pearl Harbor attack, as an aircraft mechanic. This initial combat by P-40s with the Japanese A6M Zero fighter was an eye opener for the American volunteer pilots. They flew the Curtiss P-40 single engine fighter against the Zero. The P-40 was a heavy, structurally rugged aircraft with self-sealing fuel tanks, and armor surrounding the pilot. It was

pitted against the extremely lightweight Zero, which did not have self-sealing fuel tanks or armor. The Japanese considered their pilots expendable.

The P-40 desperately needed a turbo supercharger to regain engine performance as the air density decreased with altitude. However, due to limitations on scarce materials, the turbo superchargers were prioritized for the bomber fleet. The American pilots learned the hard way not to dog-fight with the lighter and more maneuverable Zero, but rather to climb above the Zero and make diving attacks, preferably out of the sun, followed by another climb. The Zero flamed easily when the fuel tanks or fuel or oil lines were hit, and could not take much combat structural damage.

The Army Air Corps desperately wanted to get its hands on a Zero to fly against the U.S. latest fighters to develop combat tactics. The opportunity came when an American pilot in a P-40 spotted a downed Zero. Gerhard Neumann was flown to the crash site and almost single handedly put the Zero together so it could be recovered. This was the second Zero captured by U.S. forces, the first being the aircraft that crash-landed at Dutch Harbor. The latter aircraft was shipped to Wright Patterson Army Air Corps Base, and was flown against the U.S. fighters of the day in mock combat. Gerhard's efforts were so appreciated that, after the war, he was made a U.S. Citizen by act of Congress. He subsequently came to GE Aircraft Engines and worked his way up to Vice President.

When Gerhard was in middle engineering management, in Evendale, he was responsible for the design of the variable compressor stator system, which allowed the jet engine to respond more rapidly to sudden throttle movements without entering compressor stall.

Gerhard was an electrifying and motivational speaker. I sat in the cubicle occupied by the on-site representative from the McDonald Douglas Co., discussing something related to my job. All of a sudden, Gerhard walked into the office, and in his rapid fire manner asked what we were working on, etc., and made both of us feel important. The shop workers worshipped him, because he never was afraid to get his hands dirty, and talked their language. I must add, <u>none</u> of the successor Vice Presidents of Aircraft Engines, during my 31 years, mixed it up with shop people like Gerhard.

Captured Japanese Zero in US Army Air Corps Insignia
Internet photo Ref: http://en.wikipedia.org/wiki/File:A6M5_TAIC.jpg

New Bomber Engine Program
It is interesting to note in those days, the company was run by engineers who had worked their way up the chain, out of engineering into management. More often than not, GE bid a proposal just to deny the award to the competition. Subsequently, GE always found new applications for the engine to make the effort worthwhile. Today, the company is run by hot-shot MBA graduates of Harvard Business School, MIT Sloan School, or Stanford, and no risks are taken.

A classic (in my mind) example of this occurred while I was in Evendale, during the dark layoff days. The U.S. Air Force requested GE bid on a follow-on bomber demonstrator program, after the B-70 was cancelled. I guess GE had had it with bomber programs by then, and especially since this one entailed a small number of engines. GE decided to *No-Bid* the contract. The Air Force twisted some arms and GE was forced to bid, very reluctantly.

GE actually won the contract (which led to the development of the F-101 engine for the B-1 bomber), much to the company's surprise. In today's environment, the business people would have stuck to their guns and *No-Bid* the contract. Eventually, this engine design was spun off into a commercial engine program, in partnership with SNECMA of France (A French Engine Company, now called Safran), named the CFM-56, probably the most successful turbofan engine of all time.

Presently, this engine is on the retrofitted stretched DC-8, the Boeing 737, by the thousands, as well as on numerous European Airbus aircraft models. Military applications include the retrofitted KC-135R jet tanker, Boeing E-3 AWACS, Northrop Grumman J-STARS, and the Boeing RC-135. Imagine if today's bean counters from Harvard, Sloan School or Stanford had no-bid the contract!

Job Interviews in GE Lynn for New GE12 Demonstrator Engine
I somehow muddled through two years at the Evendale plant, actually hating to come to work every day. What saved my sanity was the great flying I experienced in the Ohio Air National Guard. Suddenly, GE Lynn was awarded a contract from the U.S. Army for a new helicopter engine demonstrator program, the GE12, and GE Lynn was hiring. I explained my dilemma to my boss and he was sympathetic to my desire to move back east. With his help, I secured a series of interviews in Lynn for the GE12 demonstrator program. The jobs were in Design Engineering in Engine Controls and Lube Systems, as well as a Project job. Sad to say, I was not selected for any of these jobs, and I went back to depressed mode in Cincinnati.

Outdoor Engine Test Site at Peebles, Ohio
The CF6-6 engine was still in development, prior to undergoing FAA certification testing. A development engine was undergoing test at the GE outdoor test site in rural southeastern Ohio, near a town called Peebles. I was tasked to visit this site to work on a problem. The two-hour drive was depressing, as the two-lane road wound through Appalachia, and the poverty was astounding. Some of the shacks occupied by residents near the road reminded me of photos I had seen from the Great Depression.

Well, lunchtime came, and I was told the nearest restaurant was in a small town called Hillsboro. I was able to find the town and the restaurant, which was set back from a small gas station at the intersection of two small roads. What happened next reminded me of *Ma and Pa Kettle*, from the early TV series.

Upon entering the restaurant, I observed the cook and co-owner, a heavyset woman in a housedress, slaving over an open stove. Her husband, and co-owner, was dressed in full length coveralls, and sat at the rear of the room, showing off his new hunting rifle to a friend. I sat down at a table. *Ma* came over and said, "We have meat loaf." That was the *menu*. I decided to order meat loaf.

After the meal, *Ma* asked if I wanted desert. I said "Yes," and she came over with a pint cardboard container of ice cream, a spoon and a plate, whereupon she scooped some ice cream onto the plate. I still laugh over this to this day. Hillsboro was known as the birthplace for Roy Rogers, and there was a small storefront museum dedicated to him there.

Abortive Attempt to Return to Hamilton Standard

I stayed in contact with an engineer and friend at Hamilton Standard who had been in the same engineering group. He received the only raise in engineering when I left, a raise slated for me! He came to Clinton County Air Reserve Base, where I was flying out of, to direct a retrofit of the Air Force Reserve C-119 aircraft propeller, from the Hamilton Standard hollow steel four-bladed propellers to a three-bladed solid aluminum model more available in the Air Force inventory, surplus. I flew one Friday night, and arranged to meet with my friend after I landed.

I told him I was disappointed in the way my job was going at GE. He suggested I write the chief engineer at Hamilton Standard, asking to come back. I wrote the letter, and the chief engineer called me all excited about getting me back. Shortly thereafter, I received call from the personnel representative, who told me there was a hiring freeze, and there would not be a job offer, and he wished me well.

Joining GE, Lynn as a J85 External Configurations Design Engineer

A year later, as I was about to depart with my family for a Thanksgiving vacation back to the east coast, I took a chance and called the personnel officer in Lynn with whom I had dealt previously. I told him I was leaving on a Wednesday, driving east, and would be available if there was any interest for an interview in the Lynn plant. He told me to call him on Monday, and he would circulate my resume again. When I called, he told me he had set up an interview with J85 Mechanical Systems Design. I took the interview, and it went well. The hiring manager told me he was going to make me an offer.

I was scheduled to clear through the personnel officer's boss before I left the plant. As I sat in his foyer, looking at him from afar in his *Taj Mahal* office, I heard him shooting the bull with someone on the telephone for an extended length of time, making me cool my heels, with his feet up on his desk. I didn't appreciate this, but I was in no position to complain. When I finally sat down with him, he asked me how it went. I told him the manager said he was going to make me an offer. I must have pushed it too far, because I asked him how long it would take to extend an offer. He jumped down my throat by saying at the top of his voice, **"HE SAID HE WAS GOING TO MAKE YOU AN OFFER DIDN"T HE?"**

I never forgot that humiliating incident, and had several other opportunities during my 28 years in Lynn to observe this arrogant person. Suffice to say, I <u>did</u> get the offer, and came to Lynn. I carried a grudge against this arrogant person, for many years, but I have mellowed, as he has died, and I am still alive!

GE Lynn had responsibility for the smaller engine models compared to Evendale. There was a greater profit margin in the large engines, so the staffing in Evendale was greater than in Lynn. Consequently, Lynn had to give its engineers more responsibility than the Evendale engineers. This was fine with me, as I thrived in that situation.

History of the J85 Engine

The lineage of the J85 engine began in 1956, when the Air Force selected GE to design an engine for the GAM-72 Quail. This was a subsonic jet-powered air-launched decoy cruise missile designed to be launched from a B-52 strategic bomber. The engine was designated as the J85-GE-7, with a static thrust of 2450 lbs., without afterburner. From this humble beginning, the J85 models proliferated, both militarily and commercially.

ADM20 Quail Decoy
Internet Photo Ref: wikipedia.org

Job Responsibilities as a J85 Mechanical Design Engineer

My job was to provide manufacturing support for all J85 engine models, as well as to support the fleet in the field, with redesigns to service-revealed deficiencies. We had a robust Component Improvement Program (CIP), which was funded by all the customers. At conferences held twice per year, the customers decided which problems to fund at GE. Then, in my area, I worked on field fixes for problems, and presented my progress at the conferences. During this time, I worked my way up to Senior Engineer.

Eight Stage Engine (T-38 and F5A/B/C/D Applications) Variable Exhaust Actuator Failures

About a year after I joined the J85 design team, the Air Force approved an engineering change proposal (ECP) to incorporate engine stage five cooling air to the variable exhaust nozzle mechanical actuators. I was unaware this change was in the works, as it had been designed by my predecessors in J85 External Configuration Design.

The objective was to reduce the high failure rate of the mechanical actuators positioning the Variable Exhaust Nozzle (VEN). The 8-stage engine incorporated a VEN power unit driven off the engine accessory gearbox. This unit drove three flexible shafts that worked the VEN actuators through a worm gear mechanism. This worm gear was packed in high temperature grease for lubrication, but the hot afterburner casing the actuators mounted on ran at 1100 degrees F, which was higher than the grease could withstand. The solution was to pipe compressor stage five bleed air back to the actuators to *cool* the actuators. This all happened before my time.

When I found out about the ECP approval, I directed our model shop to build a mockup of the piping system and mount it on the full scale mockup of the engine, in order to become familiar with the change. It immediately became clear the piping system did not fit on the engine, due to a drafting layout error not caught by my predecessor. I notified my management and immediately embarked on a program to correct the drafting layout error.

I earned a lot of recognition within my organization for successfully straightening out this situation, in minimum time, and the kits of parts that hit the field were correct for the engine. Naturally, drafting was protected for generating the error in the first place, as drafting was unionized, and they were not responsible for their work!

Family of J85 Engine Models
The table below outlines the family of J85 engines. The term wet or dry refers to whether the engine had an afterburner or not. The dry version did not have an afterburner. The CJ610 is the commercial designation for the J85 without the afterburner. The CF700 was a CJ610 with an aft mounted fan. The CJ610 engines were on the original Learjet business aircraft (later Learjet models had a competitor's engine.) The CF700 powered the Falcon 20 business jet, which launched Federal Express into the business of overnight

package delivery.

J85-GE-5 Powered the T-38 Pilot Trainer Aircraft
Internet Photo Ref: http://en.wikipedia.org/wiki/File:T-38_051017-F-0000S-002.jpg

Unions!

I solved a vibration problem with a tube and bracket support system on the J85-GE-21 engine. As was the custom, I required a fit check of the new bracket on a production-configured J85 engine before proposing the change for production. I made the necessary arrangements with the foreman in engine development assembly. He called over his most experienced assembler, George (who was old enough to be my father). This was not an easy assembly job, as the bracket was located in a congested area, and was mounted off the accessory gearbox split line. If a nut was dropped into the accessory gearbox area, it was a time consuming task to find and remove the loose part.

I showed George what I wanted done. He said he needed to get his tools. The foreman left for other tasks, and I waited for George. About 45 minutes later, George had not returned, but the foreman returned to see how the job was going. I explained George had gone for his tools and had not returned. The foreman took off in a huff to find out what happened.

He returned five minutes later, with steam blowing out of his ears. He said, "George felt we were picking on him for giving him all the hard jobs, and went home!" The foreman was screaming he was going to issue a "Contact Report" on George, etc. The reality was there was nothing management could do with the union shop, and George got away with his childishness. Working with unionized workers was difficult most times, and impossible occasionally. If union contract negotiations were coming up, the situation became untenable. Luckily, my need to work with unionized help was minimal.

Family of J-85 engine models

Engine Model		Application	Quantity per A/C
J85-GE-7	Dry	GAM-72 Quail	1
J85-GE-4	Dry	T-2C Talon,	2
J85-GE-17	Dry	C-123K, Booster	2
J85-GE-17	Dry	C-119K, Booster	2
J85-17A	Dry	A-37B, Dragonfly	2
J85-GE-5	Wet	T-38A Pilot Trainer	2
J85-GE-13	Wet	F-5A and B	2
J85-CAN-15	Wet	CF-5 Fighter	2
J85-GE-21	Wet	F5E/F Fighter	2
CJ610	Dry	Lear Business Jet	2
CJ610	Dry	Hanza Business Jet	2
CF700	Dry	Fan Jet Falcon 20	2

CF700   Dry   Sabreliner 75A/80A   2

Items of Interest (IOI) and My Subsection Manager

Everyone was strongly encouraged to submit a hand-written daily IOI to their supervisor. He then compiled all the IOIs in his organization and submitted a summary to his manager, and on up the line. This was a way to keep upper management informed as to the progress of the various programs.

My subsection manager was a nitpicker. He liked to rewrite all the submitted IOIs, using a red pencil. I decided I was going to write an IOI so perfect he would not be able to find any fault, and he would be forced to submit it as I had written. I could see a glancing view into his office from my desk. At lunchtime, when he usually reviewed and rewrote the IOIs, I observed him reading my IOI. I could almost see the gears turning in his head as he struggled to find something wrong with my submittal.

Finally, he came to my desk with my IOI, and told me next time, use a number two pencil. I silently roared with laughter, as this was the only nit he could find. Subsequently, he had the whole group to his house for a Christmas party. His wife was a public school teacher, and she bossed him around all evening. I then understood what made him tick. He was so frustrated with being bossed around, he had to get his licks in at the office, especially with respect to the red pencil.

Jet Engine Design Background

In the design of hardware for a jet engine, the steady state loads on a part are fairly easy to calculate, and factor into the design. The unknown is the vibratory loading. Every part, with its mounting system, has a natural frequency. Engineers try to design a system such that the natural frequency is well above the engine normal operating range. For clarification, a very small unbalance in the engine rotors results in a 1XRPM vibration. For the J85-GE-21 engine, the top speed is 16,500 RPM. A small unbalance at 16,500 RPM results in a vibration excitation of 16,500RPM/60 Cycles per Minute, or 275 HZ (Cycles per Second). If a part has a natural frequency of 275 Hz, the engine at top speed could (but not necessarily) excite it. This "tuning fork" effect causes the part to go into resonance, and usually fail in vibration.

Most hardware returned from the field for engineering investigation failed under vibratory loading. The engineering tools available to the engineer improved considerably from the time I moved to Lynn until the time I retired. Now, GE has a Unigraphics computer program facilitating the calculation of the vibratory loading during the design phase, making for an efficient design process.

There is another mode causing jet engine parts to fail in vibration, other than by resonance at the natural frequency match with the engine speed. If there is oil in the engine rotor, there will be a vibration set up that will not be exactly proportional to engine speed. Also, if a tube or manifold assembly is connected between one accessory and another accessory, driven off the accessory gearbox, the part could experience a vibration excitation proportional to the gear passing frequency of one or both accessories. An example of this could be a manifold connecting a gear driven fuel pump and a gear driven fuel control. Both accessories are mounted on the accessory gearbox, and gearing in the gearbox drives both.

One other vibration occurs when an engine is shut down and restarted while still hot. The center shaft connecting the power turbine to the fan, for example, will be hotter on the upper fibers of the shaft than the lower fibers, due to heat conduction upwards after engine shutdown. This makes the upper fibers longer, due to thermal expansion than the lower fibers, resulting in a "bowed rotor." If this condition exists, then a 1/rev vibration could be set up, and could excite engine hardware. This is normally avoided by running the engine at idle for a finite amount of time, say two minutes, in order to cool it down, before shutdown.

Jet Engine External Hardware Material Selection Rationale
External hardware on a jet engine cannot be life limited. Stated another way, all the external parts must have infinite life as a design

goal. Many fires occurring in a jet engine installation are caused by flammable fluid leakage at external tubing end fittings, spraying onto hot engine casings and/or frames, due to insufficient coupling torque. Another mechanism causing a fire in an engine is flammable fluid leaking from a tube that developed a crack under vibratory loading, onto hot surfaces. This means there cannot be the smallest defect in the material of the parts or welds propagating into a crack under vibratory loading.

As a start, all tubing and brackets must be designed using corrosion resistant steel. Tubes actually are made of SAE 321 stainless grade of steel, and brackets are made of INCO 625 (Inconel is used on brackets because the fatigue strength is nearly double that of SAE 321 stainless steel.)

Commercial engines generally have an insulation blanket installed over hot casings, or have sufficient ventilation airflow over the engine to preclude leaking flammable fluid from igniting. As a general rule, a minimum of six air changes per minute in a compartment is usually sufficient to preclude ignition of leaking flammable fluids onto hot casings. Military engines generally do not have insulation blankets on hot casings. The military accepts the risk of leaking flammable fluid impinging on a hot casing. Military pilots are considered more expendable!

Flexible hoses are generally designed using a Teflon inner core surrounded by a stainless steel wire braid, which is in turn surrounded by a fire sleeve (also acting as a chafe guard).

GE Aircraft Engines Vice-President Gerhard Neumann Walks the Shop

GE Aircraft Engines Vice-President Gerhard Neumann was based in the Lynn plant, because he loved to sail his boat out of Marblehead Harbor. Accordingly, the headquarters for GE Aircraft Engines was in Lynn, Massachusetts. Gerhard had the habit of coming into the plant in the middle of the night, with his German shepherd dog, to find out what was really going on, as opposed to laundered news from his staff.

On one occasion, on third shift, in the wee hours of the night, he wandered into Building 29, where a sole engine assembler was working putting together a J85 engine. All the lights were out in the building, except the one directly over this job. Gerhard asked the assembler how it was going, and was told the truth; he couldn't see what he was working on, because the lighting was so poor. Gerhard immediately called the head of manufacturing, in front of the assembler, and told him to get his ass into the plant immediately to turn on the lights.

The next day, everybody throughout the plant, especially the hourly work force, knew how Gerhard stuck up for the lowly engine assembler! Unfortunately, after Gerhard retired, we never had another vice-president who related so well to the hourly work force. Furthermore, after Gerhard retired, the headquarters for GE Aircraft Engines was moved to Evendale. Instead of having two co-equal chains of command reporting up to the top, one in Lynn and one in Evendale, Lynn became a satellite of Cincinnati. This reorganization was a morale killer for engineering and management in Lynn, as they became subordinate to Cincinnati.

Eventually, senior levels of engineering and management in Lynn either physically moved to Cincinnati, or simply gave up fighting for what was right for Lynn engines and caved to the Ohio mafia. From this point onward, any promotions required duty in Ohio, where visibility to upper management was easier.

Open Job Posting System
GE implemented an open-job posting system. This change was supposed to make every open job available for everyone to see.

An individual could nominate himself if he felt he was qualified. This worked reasonably well for the lower to mid-level jobs, but as the jobs got to higher levels, management or personnel meddling were common. These later jobs were referred to as *Bag Jobs*, as the job was in the bag.

Saab SF340/CT-7 Propeller Installation Engineer

The aforementioned GE12 helicopter demonstrator program ran its course, and GE captured the natural follow-on program to convert the engine into the T700 production engine for the Army Blackhawk Helicopter. After this engine went into production, GE successfully promoted it into a commercial turboprop engine (the CT7 engine) for the Saab SF340 twin-engine commuter aircraft. Hamilton Standard designed the propeller reduction gearbox and the propeller (Dowty Rotol in England offered an alternative propeller.)

An open job posting appeared for a propeller installation engineer reporting to the commercial project organization. This was right up my alley, and I thought this was my ticket into management. I contacted the personnel representative for the hiring manager and explained I had experience as a propeller installation engineer at Hamilton Standard before I came to work for GE, and nobody was aware of this fact. She became excited and told me she would pass this on to the hiring manager. She called me back, at home on vacation, and told me I had a luncheon interview with the hiring manager.

I came in on my own time, for a noon interview. Not only did I not get a lunch, I did not get the job. The hiring manager gave the job to a *Hot Shot* who returned with his MBA from either MIT Sloan School or Stanford, and the fact he did not have any experience was superfluous; he would learn on the job!

This truly was a *Bag Job*. I was so upset I just hammered nails into wooden studs when I got home. I couldn't even get a position when I was eminently qualified. I couldn't have been more furious! I insisted on feedback as to why I did not get the job, I was told my statement on the application, I was the only person in GE with that kind of experience, reflected a possible arrogance that could be a problem down the road. They had to say something. I'm sure they had to dig for that one!

Saab SF340 powered by the GE CT7 engine
Internet Photo Ref: commons.wikimedia.com

T700/CT7 Family of Engines

While I am on the subject, GE did a good job of spinning the GE12 development experience into a successful series of turboshaft and turboprop aircraft installations, as follows:

T700/CT7 Family of Engines

Application	Type Aircraft
Sikorsky UH-60 Blackhawk	Helicopter
Army, Navy, Air Force, Coast Guard	
AH-64A/D Apache Gunship	Helicopter
Kaman SH-2G	Helicopter

Bell AH-1W/AH-1Z		Helicopter
Bell UH-1Y, US Marine Upgrade		Helicopter
Canadian Search and Rescue CHS		Helicopter
EH101		Helicopter
Saab SF340	Regional Airliner,	Turboprop
Bell 214ST		Helicopter
Sikorsky S-70		Helicopter
EADS CASA CN-235,		Turboprop

A couple of years later, there appeared on the open job posting a job for a CF34 Installation Engineer, reporting to the CF34 Project. I was interested in this job, and was awarded an interview. The hiring manager showed me a document he had obtained from GE Evendale. He needed to have the successful candidate generate a similar plan for the CF34 engine. What he showed me was a hand written CF6 Installation Approval Plan. **IT WAS IN MY HANDWRITING, FROM WHEN I WAS WORKING IN EVENDALE!** I did not get this job either, because I didn't have systems and engine performance experience. At that point, I knew the cards were stacked against me, and I stopped trying to work within the GE system to move up in management. I decided to stay in design engineering and work my way up as high as I could.

Senior Engineer Job Title

After my first five years working at GE, Lynn, in J85 Mechanical Systems Design, I was awarded the title of Senior Engineer. I was anxious to attain this rating, because I could command a cubicle all to myself, as opposed to working in the open bull pen. My annual performance appraisals were outstanding and I couldn't understand why I was not selected for any of the open jobs I had applied for. I set up a meeting with my personnel representative and my manager, and laid it on the line. Either I was being lied-to about my performance, or there was something going on I was not aware of, but I should have been successful in getting at least one of the jobs I applied for!

After some post interview huddling by personnel and my manager, I was offered a job (that I had not applied for) in engineering program control for the T700 Turboprop/Turboshaft engine line. The applications for this engine were the Army Blackhawk helicopter (U.S. Navy, Air Force, Coast Guard, and Customs Service have versions), more advanced helicopter applications, and the Hughes AH64 attack helicopter. Also, there were a few turboprop versions of this engine.

Unfortunately, it soon became apparent this job had too much politics associated with it and I was never going to be successful. I hated the constant confrontational situations I was being thrust into, and I was not happy. Dealing with some of the "prima donna" managers, one would think I was working for the competition as opposed to being on the same team.

Shortly thereafter, I was selected to support a proposal effort to secure a demonstrator contract for a 5000 SHP engine, the GE27 engine, for a follow-on helicopter. When this effort finished, after about four months, I did not want to go back to program control. Luckily, a classified program was starting up, and I was able to move onto this program in my older specialty as a senior engineer in external configuration design.

Flight Operations Engineering Offer Too Late
Ironically, I received a telephone call, out of the blue, from the manager of the flight operations engineering organization in Evendale. I had already been in Lynn for several years, but he said he was finally able to hire, and wanted to know if I was interested in coming back to Evendale. I politely declined, as my goal was always to move to Lynn, and I did not want to go back to Cincinnati, Ohio.

Engineer, Set Refueling Power

Perhaps this was a career mistake.

Sikorski UH-60

AH-64 Apache

Kaman SH-2G

Bell AH-1

Bell 214ST

EH-101

Casa CN-235

Saab SF-340

Over 11,000 engines fielded in 25 types of aircraft. The engine is still in large-scale production in GE Lynn.

Internet Photos Ref; Wikipedia.org

Classified Engine Project

My responsibility while working on this classified project was to design an enclosure for the accessory gearbox, its gear driven components and piping. It was a fun job, as far as the work was concerned. We were in a locked room at a remote location. We needed to enter a combination cipher in order to gain entrance, and this cipher changed at random times, so we had to memorize it.

I worked for a subsection manager who was an excellent technical resource, but who was a poor manager of people. He used his underlings as runners to run and collect information for him, while he did all the work himself. He had a constant need to demonstrate how smart he was, even relative to his peers.

I was assigned the task of proposing features for lifting and supporting the engine in a transportation dolly. He told me to look at all the Lynn engines and to travel to Evendale to look at their engines. Then I was to make a proposal for our engine to incorporate sufficient features to accomplish the lift and support functions.

At the conclusion of the assigned two-week study, I sketched up my proposal, and requested time to review it with him. He would not meet with me, but, rather, told me to call a meeting of all the subsection managers in our conference room, and to make a presentation to the whole group. As I was making my presentation, my subsection manager sat in the back of the room and tore my concept to pieces. My own manager threw verbal bricks at me. He thought he was so smart in his critique, but in reality, he made himself look like an asshole, for not reviewing the concept first privately.

Sometime later, a high-level design review was held with the GE consulting engineering staff. I sat in on this review as an observer, and discovered the reason for my manager's constant frustration. I noticed most of the consulting engineers had the MIT college ring on their finger, as did my manager, who was about the same age.

My manager was never promoted any further than subsection

manager, whereas all the others were much more advanced in their careers. My boss just never figured out his abrasiveness had kept him back in his career.

I had had it. I wrote a letter to Tom Past, the engineering manager, and outlined my predicament working for an abrasive person, and how I was not going to take it anymore. I suggested my operation be divided up, and my function as a configurations design engineer be put under another manager, while my current manager retained the installation engineering function. I submitted my letter to the engineering manager's secretary on a Friday, while he was on a business trip. I knew he would be in the office on Saturday to catch up on his work, and I expected to get a rise out of him on Monday. Monday came and went, and there was no indication my letter was read.

As I returned from lunch on Tuesday, I was notified Tom wanted to see me. After he closed the door, he explained he had read my letter, and agreed with me. I was in an untenable situation. He said my manager had a bad habit of "Job Over-Specification." He agreed to move me out of that organization and under another more reasonable manager. But, he said, he had to make it sound as if it was the idea of my abrasive manager. It all worked out in the end. Interestingly, Tom went on to become a department general manager, and, eventually, a vice president in the Industrial Gas Turbine Division in Schenectady, New York.

As an aside, working on the same engine project, but as a system and performance engineering technician, was a young woman who, a few years later, was a passenger on the Boston to Los Angeles flight that crashed into the World Trade Center on 9/11/01. Although I didn't work directly with her, we were in a small tightly integrated engineering organization, and I knew her.

It was somewhat humorous, when my specialty had a change in management, and my new manager had to wait for several months to obtain a security clearance. He could not be told what we were working on, and had to sit outside of our locked working area.

After a couple of years, the project was cancelled and we all had to scramble for jobs. The thing I found out the hard way was when you worked on a classified project, off site, away from the mainstream of everyday contact with fellow engineers, you became forgotten.

While I worked on this classified engine project, the company came out with a push to reduce the number of sick days being used by exempt (salary) workers. We were all entitled to five sick days per year, and if they were not used, they were lost at the end of the year, rather than being rolled over. Engineering as a whole was good about only using sick days when sickness actually occurred. However, drafting, which was unionized in Lynn, always considered sick days as extra vacation days, and used all the days.

Our *Casper Milquetoast-looking* personnel officer chaired a meeting in our conference room. He made the statement, "Some people think sick days are a benefit, they are not," whereas one of our sharp senior engineers said, "That is probably because the company discusses sick days in the benefits book!" The personnel officer blanched and didn't know what to say.

T407 Turboprop Engine
I had some difficulty in securing another position, but at the last moment I was asked to support a proposal to modify the GE27 demonstrator engine for the new U.S. Navy P7 Antisubmarine Patrol Aircraft; this was supposed to replace the worn out P3C aircraft, and was to be a turboprop! Lockheed Burbank, California was the prime contractor for this aircraft, and conducted the competition for the engine.

There was a twist. Lockheed wanted the engine manufacturer to be responsible for the entire power plant package; the engine, propeller reduction gearbox, propeller and nacelle with accessories. GE competed with Allison, who was the engine manufacturer on the original P3C aircraft. GE staffed up for this effort, and pandemonium reined. I tried to be recognized as the person most qualified to be the propeller integration engineer, but I was unsuccessful in making an appointment to talk with the department general manager. *He just would not see me*, and I was frustrated, as I knew this job was made for me with my experience. I secured this position by being persistent in calling the T407 Engineering Manager who worked under the aloof department general manager. The T407 Engineering Manager eventually let me make my case, and I got the job.

I could perform as a propeller installation engineer in my sleep, and I looked forward to the effort. I reported directly to the T407 Engineering Manager, which made me feel good. The first thing I did was conduct a competition for the propeller; the two competitors were Hamilton Standard and Dowty Rotol, of England. After generating a Request for Proposal (RFQ) and submitting it to both companies, and visiting them, I evaluated the competing proposals.

Hamilton Standard proposed their latest pitch-changing concept in a five-bladed design, whereas Dowty Rotol proposed a lightweight six-bladed design. Both proposals incorporated composite propeller blade construction. In the end, I selected the Hamilton Standard design, as they had more experience in high horsepower propeller applications, and were more responsive to the requirements of the RFP.

GE won the contract to design and develop the powerplant for this aircraft. The engine was renamed the T407. What followed was three years of intensive design coordination between Hamilton Standard and GE, and I was in the middle of it all. This had to be the best job I experienced since joining GE. I made several trips to Lockheed in California, and many trips to Connecticut to Hamilton Standard. Some of the people I worked with I had known when I worked there many years earlier.

The engine design effort was finished and the engine development phase was underway. Development engine hardware was ordered and delivered. GE purchased a complete nacelle from an aircraft

"Bone Yard" off of a retired Lockheed Electra aircraft, which was the predecessor for the P3C aircraft. The intention was to build up a complete powerplant inside the nacelle for a total systems test at the GE Peebles outdoor engine test site in rural Ohio, near the West Virginia Border, in the heart of Appalachia. I was among the group of engineers supporting the buildup of the powerplant inside the nacelle on-site. Hamilton Standard had their team of engineers and mechanics to support the propeller system buildup.

It was the end of June 1990, and my manager was also on-site, because his incentive compensation (bonus) was predicated on firing the powerplant system to at least idle power by 6/30/90. Well, it was 6/30/90, and every time the test technicians tried to fire the engine, they discovered a fault. This was normal for a new program, but my manager had his bonus at stake, so there was much more stress.

We worked all day long, from about 8:00 A.M. and it was now about 9:00 P.M., and we were still trying to fire the engine. We hadn't eaten since breakfast, so our manager sent out to town for pizzas. We got several slices of pizza and a soda, and he got a bonus of up to 50% of his salary. I think this was a fair comparison!

Anyway, we finally got the engine up to firing speed on the starter motor; the engine fired, and was accelerating up to idle, when the propeller spinner disintegrated! The spinner was made of fiberglass, as was Hamilton Standard's convention, but the fibers were incorrectly laid 90 degrees to the design intent, thereby making the spinner much less capable of reacting to centrifugal force while operating. The design was proper but the part was made wrong. I was supposed to be inside the engine test control room, but I wanted to observe the propeller coming up to speed, and was outside on the roof of the control room.

I witnessed the whole event, with fragments of the spinner being batted around by the propeller blades. We returned back to Lynn, and I worked with Hamilton Standard to find out what happened. A few days later, the U.S. Navy cancelled the whole program; I was crushed, as I was hoping I could ride this job until near retirement.

Our competitor, Allison, continued to work, on its own money, on a modernized version of the old T-56 turboprop engine in the C-130 and P3 aircraft. Allison worked with Dowty Rotol of England on their six-bladed composite propeller design and proposed a commercial contract with Lockheed Georgia for a follow on version of the C-130, called the C-130J.

GE got wind something was going on, and sent a team of engineers, myself included, to Lockheed Georgia to pitch our T407 engine. When we arrived in the conference room with Lockheed, we were immediately told Lockheed had a sole source commercial contract with Allison for this application, and they were not interested in talking to GE about this program. I blame our marketing representative for completely dropping the ball on this potential program. Allison aced GE out by about six months. The C-130J went into production for the U.S. Air Force, is currently in service around the world, and is continuing to sell well.

C-130J powered by a modernized Allison turboprop engine and a six-bladed Dowty Rotol Propeller

Ref: dyes.af.mil

Engineer, Set Refueling Power

Another Massive Layoff in Engineering
Next followed a massive layoff at GE, Lynn. It was advertised as a top to bottom review of every position and the person in the position. We were told this was to be the *Mother* of layoffs. I was nervous. The process took about three months, during which time I had absolutely <u>nothing</u> to do. Finally, the notices were made as to who was to be laid off, and I luckily made the cut. As was the GE practice, those people who were working in engineering, but who were not actual engineers, but technicians, were always the first to be laid off.

I felt particularly bad for the T407 test engineer, who was probably actually a technician. He was a talented person, with a great deal of experience in the test side of the house, particularly on T64 turboprop engines. He was less than one year short of the sanctuary of 25 years GE service, where the company found you a best offer to stay employed. This person went out the door! I ran into him a few years later in Newburyport, Massachusetts, at a theatre. He was still depressed and sour on the company.

This Lad Doesn't Know What He is Talking About
One of my engineering friends, who was now working in the preliminary design engineering group, wrote a letter to upper management. He made the case this was the time management should step up and continue to employ engineers to work on the next engine for a regional aircraft that would be economical to operate at reduced passenger counts. This letter made its way all the way up to the Vice President of Aircraft Engines, Brian Rowe (Originally from Roll Royce, a Brit) at the time. Brian made some hand written notes on the letter and sent it back down the chain, until it landed on my friend's desk. The hand written note said, "This lad doesn't know what he is talking about." Several years later, the company got into the regional jet business with one of Lynn's engines, the CF34, and it continues to be a tremendous success. I guess this lad knew what he was talking about after all. He was just ahead of his time.

Engineering Reorganization

Engineering was completely reorganized into a matrix organization, whereas each group of engineers supported all the engine lines rather than be assigned to just one. I was assigned to an organization comprising both installations engineering and external configuration design. With the help of my new manager, I carved out a job as the specialist in Lynn for engine fire safety, supporting all engine lines. This was a fun job, as I wrote a revision to the Design Practice for Installed Engine Fire Safety.

While performing as an engine fire safety specialist, I was assigned to work on the impact of incorporating an FAA Advisory Circular having to do with testing aircraft engine parts for fire proofness and fire resistance. A part (tube or flexible hose) was considered fireproof if it could withstand a 2000-degree F flame for 15 minutes without failing such that leaking flammable fluid would feed the fire. Likewise, fire resistance was withstanding a 2000-degree F flame for five minutes without feeding the fire via leakage.

By definition, a fuel carrying tube or hose needed to be fire resistant (five minutes), as it was assumed a fuel fire could be starved for fuel within five minutes, by virtue of shutting down the engine and fuel shutoff valve once the fire was recognized in the cockpit. An oil fire, on the other hand, was fed from the oil tank on the engine, and there was no way to shut off the oil supply from an oil tank, until the whole oil supply was depleted.

I was involved with a team of engineers to prove the existing production CF34 engine firewall seals could withstand 15 minutes of flame at 2000 degrees. Our testing proved the production seals could not pass the new test. We devised a new concept for a firewall seal to pass the 15-minute fire test. I pushed for the submittal of a patent application with the U.S. Patent Office, and the patent was eventually awarded, with all of our names on it. Of course, the company owned the patent, as signing a patent disclosure form was a condition of employment when coming aboard to work for GE. However, the company generously awarded all of us with a $200 *Good Inventor* payment! Wow, where do I spend it first?

This job lasted for a couple of years, until another reorganization took place. I was now becoming a fairly senior level engineer in my specialty. Whole layers of management were eliminated, and responsibility was pushed down into my level for the direction of technical effort. I was assigned responsibility for **all** the external configuration design effort for the F404 model line, and reported to subsection level in management. I provided technical direction to several engineers working on the F404 engine program. Of course, there was no increase in compensation for the increased responsibility!

Six Sigma Push

The GE President and Chief Executive Officer, Jack Welch, became enthralled with Six Sigma. This was a statistical means of reducing the error rate for any repetitive process to less than 3.4 defects per million opportunities.

It was a process of gathering the measured data, analyzing the data, finding out the reason for the error rate, and making changes to reduce the error rate. Jack Welch rammed this concept down the throats of everyone in the company. He was so enthralled with it he made the statement no one will be promoted without becoming an advocate, or black belt, for Six Sigma. Everyone had to become trained, and awarded a green belt title, upon completion of the training. Selected individuals were trained in a follow-on project to become mentors, and were titled Black Belts. No one was to be promoted without becoming a black belt going forward and spending at least one assignment mentoring some organization within the company.

I understood the merits of applying Six Sigma to any process generating lots of data, such as in manufacturing, where all the tolerance dimensions on the drawing for the part must be met. Say if a part had 20 dimensions on the drawing, and 100 parts were made in one run, there were 100 sets of data on the 20 dimensions. <u>BUT</u>, in engineering, only a few sets of development hardware were ever made, usually to qualify a new design or a redesign. This definitely didn't lend itself to Six Sigma processes. The concept was rammed down our throat. I actually observed engineers with considerable service leaving the company, because they did not want to be involved in this statistical mania going on. One of these engineers lives in my neighborhood, and works for a job shop servicing GE, locally. We all laughed at the GE statement only black belts would be promoted. I learned some time ago, when GE promotes a policy, GE means it!

There are currently many senior people in engineering (senior engineers, principal engineers, and senior staff engineers) who are stagnant in their career because they do not want to participate in the black belt process; they simply are waiting it out until retirement!

The F404 Engine Applications

The F404 engine (with afterburner) powered the McDonald Douglas (now Boeing) F-18A/B and C/D Hornet models for the US Navy, Marine Corps, and miscellaneous foreign countries, such as Canada, Australia, and Finland among others. The engine, without afterburner, powered the Lockheed F-117A Stealth Fighter. An afterburning model powered the Hindustan Aeronautics Limited Tejas indigenous fighter for India.

In addition, the Singapore Aircraft Industries retrofitted the A4 to the A-4S Super Hawk, which resulted from replacing the obsolete Westinghouse turbojet with the F404-GE-100D dry engine. The F404-RM12 engine (with afterburner), in partnership with Volvo Aero Corp., powered the Saab/BAE JAS 39 Grippen in service in Sweden and South Africa.

The F404-GE-102 engine (with afterburner) powered the South Korean T-50 and A-50, pilot trainer and lead-in combat fighter. Also, there was an Industrial version of the engine, the LM1600, driving a compressor to pump natural gas across the Canadian pipelines as well as a few marine applications.

Engineer, Set Refueling Power

F404 Engine Applications

Engine Model	Wet/Dry	Application	Qty/Aircraft
F-404-GE-400	Wet	F/A-18A/B Hornet	2
F-404-GE-402	Wet	F/A-18C/D Hornet	2
F-404-GE-100D	Dry	A-4S Super Hawk	1
F-404-RM12	Wet	Saab/BAE Grippen	1
F-404-GE-102	Wet	T-50/A-50	1
F-404-GE-IN20	Wet	Tejas (India)	1
F-404-GE-F1D2	Dry	F-117A Stealth	2

Failure of P3 Manifold

I had some interesting problems to solve on parts failed in service due to vibration, and I received good of visibility to upper management.

One I was particularly proud of was a tube manifold carrying a signal pressure (compressor discharge pressure, or P3) to the main fuel control. The part had a significant failure rate in-service with the U.S. Navy on the F/A-18A/B/C/D aircraft in aircraft carrier service. Failure resulted in the engine decelerating to a sub-idle rpm (power).

I arranged for an analytical stress analysis leading to an instrumented engine test. The results confirmed the part had a natural frequency excited by the engine near top speed, and the vibration stress level made the part life limited. A redesign solved the problem. My job required I brief the F404 Project Manager biweekly on the progress of the failure investigation and redesign. This F404 Project Manager was destined for eventual vice-president (where he is currently), and I made a good impression.

New Manager from Hell

Eventually, my administrative supervisor in Lynn left the company and the engineering mafia in Evendale anointed an Evendale senior staff engineer to take his place. In order to make the deal even sweeter, the new supervisor worked out a deal to be based at the GE tube fabrication facility in Hooksett, New Hampshire, 90 miles away, where he would not be subject to Massachusetts state income tax.

To be fair, this individual <u>was</u> an excellent technical resource, but was an atrocious manager of people. He spent most of his time in New Hampshire, and drove to Lynn once a week for a whirlwind tour of his troops. Typically, he sailed into my office and asked, "How is it going?" As soon as I said anything, I got a five-minute lecture on how the business needs to do this and that! It repeated the next week.

I effectively managed the F404 external configurations design effort with my assigned team of younger engineers. I made all the decisions, both technical and managerial, to keep my effort moving forward. I **<u>NEVER</u>** had input from my administrative supervisor, in Hooksett. I probably would have been successful as a small business owner, as I performed most of the functions of a business owner; recruitment, daily work direction, managing the budget, etc.

I Went to War with My Management

When the time came for my salary action, after being strung out to 18 months since the last action, I was told my raise was to be 1.8%. I was in my last years before retirement, and the company figured it could get away with screwing me! I asked my administrative supervisor if he solicited input from the organization I supported (The F404 Project) on a full time basis, as to my effectiveness. He said he had. I knew he was lying, and went behind his back and made direct inquires as to whether he had made this solicitation. I confirmed my manager was indeed lying. I then went to the Lynn Ombudsman with a complaint a rogue manager was managing me. The ombudsman told me to document it all in writing, which I did. He submitted it up the chain.

Engineer, Set Refueling Power

Our Evendale-based human resources manager called me and tried to arrange a three-way meeting with my supervisor, him and me. I would have none of it. My human resources manager was upset with my call. My complaint worked its way through the system, until, one day, my rogue supervisor came into my office and closed the door. I instinctively clenched my fists in anticipation of a verbal fight.

He told me **_He_** had made a recommendation to change the salary action to 5.8%. Again, I knew he was lying. I found out the F404 Project Manager I previously briefed on a biweekly basis on the above problem resolution was now located in Evendale with a promotion to department general manager (one step below vice-president) in charge of my organization. It was he who had intimate knowledge of my work and changed the salary action. It wasn't much later my rogue manager was transferred back to Evendale. Good riddance to bad rubbish!

Another incident happened before my rogue supervisor was transferred. My team worked to resolve a problem with the fuel recycling system on the F404 powered F-117A Stealth Fighter. Normally, the F404 main fuel manifold is purged of fuel upon engine shut-down by draining onto the ground, to prevent the soak back temperature in the engine bay from coking the fuel in the manifold. This was a small amount of fuel drainage and was harmless.

However, because the F117A could not have this hole in the skin of the engine bay for radar reflectivity considerations, a system was designed to capture the fuel drainage in a container. The engine then sucked up this drained fuel on the subsequent engine start. Unfortunately, the system had a 1/rev natural frequency in the engine operating range, and a tube in this system failed in fatigue, thereby causing the fuel leakage. The incident occurred on the ground, but a subsequent fire caused by the leaking fuel demolished the aircraft, and the Air Force was extremely interested in the solution of this problem.

Engineer, Set Refueling Power

Singapore A-4S

Boeing F/A18

KAI T50 Golden Eagle

Saab Gripen

India Tejas

F117 Stealth Fighter

F404 Powered Aircraft
Internet Photo Ref: wikipedia.org

I assigned an engineer under me to redesign the system. It came time to perform a fit check of the redesigned system in the actual aircraft at the Lockheed facility in Palmdale, California. My engineer requested a travel approval from my rogue supervisor. Understand this supervisor had absolutely nothing to do with this engineering program up to this point; everything was managed completely by me.

Now my supervisor wanted my engineer to report back to <u>him</u> by telephone every evening on the progress of his fit check. I blew-up. If my engineer was going to call anybody, it was going to be me, for I was the only one who could help him if he had a problem with needing some more hardware or whatever. And I trusted my engineer such if he didn't need to call me, I could wait until he got home to be debriefed! I didn't score many points with my rogue supervisor.

LM1600 DLE Industrial Engine

GE secured a contract to demonstrate a Dry Low Emissions (DLE) version of the LM1600. The LM1600 was an industrial version of the F404 aircraft engine, without afterburner. It was procured by Canadian pipeline operators to drive a compressor to pump natural gas across Canada. This engine was powered by natural gas, which required a sophisticated engine control system, and an elaborate fuel manifold system feeding the natural gas to the engine. The engine was designed, hardware procured, and a demonstrator built-up. The concept was successfully demonstrated on a test stand. My team received an award on engineer's day for our effort to successfully develop this DLE version of the LM1600.

GE LM1600 Industrial Engine, Internet photo

Ref: gknaerospaceengine sstems.com

F404 for Korean Pilot Trainer and Lead-in Combat Fighter

Subsequent to this industrial engine effort, GE put together a team of engineers to bid on modifying the F404 engine for a single engine application for the South Korean Air Force, called the KTX aircraft. GE competed against Snecma of France, who bid a new engine. The GE proposal was a modification of a proven engine, whereas the Snecma engine was a paper engine (the engine didn't exist, yet) and GE won the competition.

The application had a twist. GE bid modifying the F404 engine to use a FADEC Engine Control (Full Authority Digital Engine Electronic Control). This was an electronic fuel control, without a hydro-mechanical engine control as a backup. This had not been done before (at least at GE, both Lynn and Evendale) where a backup control was not used in a single engine application. The technology had advanced to the level GE felt comfortable with using just the electronic fuel control. Also, since this was a single engine application, all the engine sensors were redundant; for example, if there was an oil pressure switch providing the FADEC with a signal, there had to be two of them, with redundant electrical harnesses etc.

A contract requirement was GE Lynn train five "Co-Development Engineers" from KAI (Korean Aerospace Industries). GE was to teach them our technology in designated areas of expertise. I inherited one of these five engineers. He was a good engineer, but his English was poor, and communication was difficult. We did our best with him, and he contributed to the designs.

Unfortunately, there were culture problems. One of the five co-development engineers was obviously the boss over the other four. The objective of these engineers was to learn (steal) as much as they could about engine design. Our objective was to keep them from learning any more than was necessary to just do their job. My co-development engineer asked me a question I knew he was not authorized to know the answer to, and I responded I could not tell him the answer.

His boss must have been on his case, because he continued to ask the same question, as if he hadn't asked it before. I continued to respond

Engineer, Set Refueling Power

I could not give him an answer. Finally, one day at quitting time, both my co-development engineer and his boss came into my office, and asked the same question. I gave the same answer. They turned around and went out. Then the boss came back in, and tossed a small box onto my desk and left. Puzzled, I opened the box to see a Samsung watch inside. I immediately turned the watch over to my supervisor and told him the story. This was the right thing to do. I was not going to compromise my 25 plus years at GE for a crummy watch!

A normal process of designing engine hardware was to hold design reviews with the chief engineer's office in Evendale. We needed to hold one of these reviews for the hardware our co-development Korean engineer designed. I told him he needed to make the presentation to the chief engineer. Kim went into shock. This was simply not done in his culture. In Korea, only the boss did the talking; if he was incorrect, nobody dared correct him. I reassured Kim I would not let him fail, and I interjected wherever I could to make him comfortable during the design review. He survived the experience, but he must have lost a few years off his life!

Kim had his family with him, and lived in a rented apartment in Salem, Massachusetts. His family did not speak English, so I did not know how they got along living in the U.S. He bought a used car, and the family did some traveling while he was here. He had to interface daily with our drafting designer, usually after lunch. After a daily brown bag lunch of Kim Chi (fermented cabbage), our drafting designer nearly passed out with the hot breath of our Korean!

I made one business trip to Korean Aircraft Industries in Sachon City, South Korea. The flight took 15 hours non-stop from New York to Seoul, Korea. I stayed overnight in the Grand Hyatt Hotel in Seoul, and flew down to Sachon City the next day. There was only one hotel in Sachon City, the Bong Dong Hotel.

Those of us who traveled to the Bong Dong Hotel wondered why we always got the same hotel room. I'll leave it to the reader to make a conclusion, but let me say we were careful what we said in the room and on the telephone. This hotel had two restaurants, one for western food and one Korean food. I naturally ate in the western food restaurant, but I simply did not recognize the shrimp cocktail or the steak I ordered. I also noticed there were not many dogs wandering around in the neighborhood.

The design of this engine model was the first for me using the Unigraphics software used for design. This made for a more accurate design. When the first engine was built-up, everything went together the first time, without being reworked. I was in heaven, and got a lot of good press at the time. This was the pinnacle of my career at GE. I was now nearing age 62, and I decided to hang it up, as I did not see another engineering program on the horizon as interesting as this one.

Request for Voluntary Lack of Work Denied

Per the GE administrative procedures, I submitted a request for a "Voluntary Lack of Work." This would have given me a lump sum payment to go out the door. My management wouldn't hear of it, and so I left anyway. Exactly 41 days later, on 9/11/01 the attack on the World Trade Center occurred. Subsequently, engineers in other groups were awarded the "Voluntary Lack of Work" I was denied. To my knowledge, only one engineer in my organization, both Lynn and Evendale, was awarded this action, ever. This man was in his late 70s in age, and had long since ceased to perform to the standard expected of us. He simply refused to retire, so they made him an offer he could not refuse. Other groups handed out "Lack of Work" notices freely, but not mine. There definitely was a disparity in how this administrative procedure was implemented, and it was not fair.

Engineer, Set Refueling Power

F404-GE-102 powered the Korean Golden Eagle
Ref: en.Wikipedia.org

Looking Back on My Career at GE

As I look back on my career at GE, I didn't advance as far, professionally, as I knew I was capable of. When I joined GE, upper engineering management consisted of people who had worked hard and eventually were promoted to their lofty levels. But everything changed very insidiously, such I could not see what was happening until long after it happened. Somewhere in the middle of my career at GE, a subtle change was implemented to promote only engineers who graduated from three named MBA programs; Harvard, MIT Sloan School and Stanford University. These engineers either left the company to pursue their MBA on their own, or were subsidized by GE to attend. MBA degrees from other institutes of higher learning just didn't count. One day, I looked around and saw the climate had definitely changed. The only exception to this was the result of affirmative action.

Another fact bothered me. When business was good, engineering didn't share in the company's good fortune. The attitude was engineers should be grateful to have a job. But when business turned, as it inevitably did, layoffs resulted, and engineering got hurt. This was a situation in which the engineer could not succeed. They were just hired help.

GE Hiring of New Engineers

GE hired only the best engineers out of college, with a 4.0 Grade Point Average (GPA), and in quantity. Since I did not hire directly out of college, but rather from another company, I was not faced with the GPA requirement for new college graduates. With all these sharp engineers vying for promotions, it was difficult to break out of the pack.

If an engineer stayed in engineering for a career, he or she attained the level of senior engineer or senior staff engineer and then stagnated. If the engineer actually obtained a doctorate in an engineering disciple, he or she rose to principal engineer. The bottom line was engineers were treated as hired help or necessary overhead. Management felt it was essential to keep engineering salaries in check, so once the engineer had, say, 10 or more years of experience, the salary progression was poor, and certainly did not keep up with inflation.

Salary Caps

I developed an empirical formula over the years to calculate the highest salary an engineer could earn. To my knowledge, no engineer earned more than twice what newly hired inexperienced engineers earned. If new engineers were offered $40,000 per year, experienced engineers never made more than $80,000 per year. One reason for this was salary progression to hire new engineer graduates, year over year, was reasonably rapid.

Experienced engineers seldom received raises in salary annually. Most of the time the salary increase cycle was 15 months, and sometimes it was 18 to 24 months. GE used the participation method to keep engineering salaries in check. In any one year, a participation factor might be 85%, along with a 15-month raise cycle. This meant only 85% of the engineers were going to get a raise, even after waiting 15 months.

The way this was implemented was to take those who normally were to receive a raise near the end of the year, and push them into the next year. This met the 85% goal, and resulted in the engineer waiting sometimes 18 months to get an increase. The amount of the raise was never factored into the raise timing interval, so it was easy to see inflation ate away at engineering salaries over time.

I looked back to my employment at Hamilton Standard in Propellers, in the late 1960s. I realize times were different, perhaps, but the company made a cost of living adjustment, annually, in addition to a salary adjustment. Also, at Thanksgiving time, the company made "Turkey" payments. In effect, they gave a bonus to the engineering workforce. The amount of the bonus was dependent on the engineer's salary level. In all my years at GE, which started immediately after I left Hamilton Standard, I never experienced a cost of living adjustment or a bonus!

In my early years, I was reasonably satisfied with my salary and the interesting work. I thought I was well ahead of my peers who graduated from college in fields not technical. However, looking back, I now know my feeling was probably true for the first half of my career. However, college graduates who were working in fields such as sales, for example, passed me big time in terms of earning power. Accordingly, I could not recommend engineering as a career path to high school graduates. The course of study is difficult, the opportunity for fun in college reduced due to study requirements, and the engineering appreciation by management when working in industry is not there.

Drive Responsibility Down to Senior Staff Engineers
After the last layoff, called a reduction in force, whole layers of management were eliminated. There were no more unit managers, and senior staff engineers were given total technical responsibility for

the product they were working on.

Senior staff engineers provided less experienced engineers guidance on all technical matters. The senior staff engineer and the engineers under him/her reported administratively to a subsection manager in all matters related to personnel, including salary. However, there was no increase in compensation given to the senior staff engineers.

They were too high in level to be paid overtime, but not high enough to be paid bonuses. So what happened was lower level engineers were paid overtime (straight time earnings were paid for hours worked in excess of 40 per week), if overtime was deemed as a requirement by management.

Sometimes, a minimum of four hours had to be worked overtime for free before straight time earnings were paid for further overtime. At the next level higher than a senior staff engineer, section level, a bonus program was in effect, so they were effectively getting the equivalent of overtime. Senior staff engineers were caught in a catch 22, so to speak. ***It is interesting; the very technical people responsible for generating a product for the company to sell and make a profit, were the very people who were taken advantage of.*** I do not want to hear lofty CEOs of large companies pontificate about the shortage of engineers being graduated from U.S. colleges and Universities. These CEOs set the tone, and the chickens came home to roost.

CHAPTER 11; AFRES AND C-123K EXPERIENCE

My Experiences in the AFRES with the C-123K Aircraft

I was anxious to relocate from Cincinnati, Ohio to my hometown of Lynn, Massachusetts and managed to secure a transfer to the GE Small Aircraft Engine facility, in Lynn, in 1973. By then I was a senior captain, and had earned my senior navigator aeronautical rating, by virtue of passing the 2000 hour flying mark and seven years rated service. I also passed the 1500 hour mark in the KC-97 aircraft, alone. I neared eligibility for promotion to major, and knew it would be difficult to secure a *slot* in an ANG or AFRES flying unit, as the *"major"* slots were reserved for captains in the unit who were promoted. There was a table of organization dictating so many lieutenants, so many captains, etc. The Air Force had secured permission to keep an officer one grade higher than the table of organization specified. In other word, a captain could occupy a lieutenant slot.

There were no openings for a navigator in the units near my new home. In frustration, I wrote letters to both my U.S. Senators asking for help. It must have worked, because out of the blue I received orders to report as a navigator to the 731st Tactical Airlift Squadron (TAS) at Westover AFB.

The 731st Tactical Airlift Squadron (TAS) was under the 901st Tactical Airlift Group (TAG). It had just moved from Hanscom AFB in Bedford, Massachusetts to Westover Air Reserve Base, 100 miles away. The reason for the move was simple economics. The Air Force had 10% of the air traffic in and out of Hanscom AFB, with the rest of the traffic consisting of light aircraft and business aircraft on the commercial side of the airfield, using the same runways. However, the Air Force paid for 100% of the airfield upkeep and navigational aids, and it had only one flying unit assigned, the AFRES C-123K equipped 731st Tactical Airlift Squadron. In the move, many AFRES crewmembers decided not to commute to the new location, 100 miles away, for one reason or another, thereby opening up a navigator *slot*.

Westover AFB previously had the C-124C aircraft assigned, which I wrote about earlier. By now, in 1973, the C-124C aircraft were retired to the boneyard at Davis Monthan AFB, and the reserve unit had the C-130 aircraft assigned. When the 731st TAS was relocated from Hanscom AFB to Westover, with their C-123K aircraft, there were two different aircraft types assigned.

C-123 Aircraft Background

The Fairchild C-123 aircraft started its lineage as a glider, the CG-20, with a shoulder wing and rear cargo ramp that lowered for ease of cargo loading. Subsequently it was made into a twin piston engine tactical airlifter (affectionately referred to as *Trash Hauler*, *Thunder Pig*, or *Polish Whisper Jet*).

The engines were Pratt & Whitney R-2800 twin row radial engines developing 2300 shaft horsepower (SHP) each. They drove a three-bladed Hamilton Standard constant speed, full feathering and reversing propeller. The early aircraft, with just the two piston engines, was considered underpowered. It was subsequently retrofitted with twin General Electric J85-GE-17 turbojet engines, one under each wing, for booster power, and was re-identified as the C-123K. Although the addition of the jet booster engines increased the payload by 1/3, they did not improve the aircraft speed significantly. However, they greatly improved the aircraft short field takeoff and climb performance and provided a safety margin not inherent in the original aircraft design.

The C-123 aircraft had no soundproofing on the inside of the fuselage. The engine noise was so loud, interphone use was the only way to communicate with the crew. Otherwise, shouting was necessary, but was minimally effective. Both the KC-97 and the C-124 were loud inside the aircraft, but both aircraft were equipped with sound absorbing material inside the fuselage.

Engineer, Set Refueling Power

The aircraft on the left is the C-123K with the 731st TAS, whereas the aircraft on the right was the 337th TAS C-130 Hercules. Photo by the late Betty O'Connell, Westover Public Affairs Officer in the 1970s.

Chase XG-20 Glider
Ref: wikiwand.com

From my personal experience in the aircraft, I witnessed great short-field takeoff performance, and initial climb, but the aircraft ran out of *poop* quickly. It cruised at 110 knots true airspeed (KTAS), which was nothing to write home about. The jets were used for takeoff, for climb, and during the airdrop. They were shut-down after level-off and after the airdrop for landing. The jets had inlet blocker doors, which were shut when the jets were not in operation. This was to preclude foreign object damage to the jets, particularly upon landing, when the propellers pulled reverse thrust and stirred up all kinds of rocks, dust, etc.

C-123K Aircraft
Nice photo, crappy aircraft
Ref: nationalmuseum.af.mil

The C-123K aircraft was relatively small compared to the C-124 and KC-97 aircraft I had previously been associated with. The maximum weight (landing gear limited) for this aircraft was 60,000 lbs., as compared to 175,000 lbs. for the KC-97G and 212,000 lbs. for the C-124   The navigator was required to plan the fuel load based on the mission, and to construct a *Howgozit* chart, plotting fuel required vs. projected distance flown, and, in flight, to plot fuel burned vs. distance actually flown. This was similar to the requirement used on the C-124 aircraft. However, this procedure was used only on cross-country flights, as local missions were only about three hours long.

The last reserve units in the Air Force inventory flying the C-123K were the 731st TAS at Westover Air Reserve Base, 758th Tactical Airlift Squadron at Pittsburgh International Airport, and both the 906/907th TAS and the 355/356th TAS at Lockbourne/Rickenbacker Air Reserve Base, Columbus, Ohio. These were the last units in the Air Force inventory with aircraft equipped with piston engines, and, as such, had the only remaining aircraft using high octane aviation gas for fuel. I was assigned to the 731st TAS at Westover Air Reserve Base.

At the time I was just happy to be able to fly again, but things changed quickly. Some of the assigned aircraft were former *Operation Ranch Hand* aircraft used in Vietnam as Agent Orange defoliant spray aircraft. One of our C-123K aircraft was nicknamed *Patches* as it had sustained more than 600 hits through its skin, in Vietnam, and had the holes patched. This is what happened when an aircraft flew low-level and slow, in a combat zone. The Agent Orange chemical had spilled over time and permeated the cargo floor of the aircraft, and could not be removed. The stink was troublesome.

C-123K Navigator's Station

The navigator station was definitely an afterthought. It consisted of a fold-down table on the backside (cargo side) of the cockpit bulkhead, in the cargo compartment, with a seat pivoted from that bulkhead, several feet in the air, on a boom. The navigational instrumentation was better than Magellan had, but not by much. It consisted of a true airspeed meter, pressure altimeter, and the J2 compass system accurate to plus or minus 2-5 degrees accuracy. This compared to the N1 compass system on both the KC-97 and C-124 with an accuracy of plus or minus ½ degree. A periscopic sextant came with the aircraft, but no autopilot, so celestial navigation accuracy was poor. (Hand flying an aircraft introduces minute changes in aircraft heading and attitude that impact the accuracy of celestial navigation.) When sighting on a celestial object (sun or star) anyone walking from the front of the cargo compartment to the rear introduced a jiggle in the aircraft, and consequently, in the bubble in the sextant. As in the C-124, the aircraft had the WWII vintage APN-9 Loran set. Also, there was a driftmeter.

There was no radar. This made navigation primarily an exercise in map reading, and precluded formation flying in weather. Whenever a formation encountered weather, it split up, with aircraft changing altitude to maintain separation in the soup. Add to this mix, the aircraft had no sound insulation, and the fuselage acted as a sound echo chamber. Ordinary conversation, other than on interphone, was impossible. Also, the aircraft was unpressurized, so, again, when it rained, we all got wet. The lack of pressurization limited the aircraft to 10,000 ft., unless the crew was on oxygen. In short, this was a basic aircraft, from a navigator's point of view.

The navigator seat could not be occupied during takeoff and landing, because it was not stressed for a crash landing, and the plane of the propellers sliced across the navigator station. In the event of a crash landing, if there was a failure of a propeller blade, it could slice across the navigator station; not a great idea. Also, from a ditching viewpoint, with a shoulder mounted wing, the aircraft theoretically settled with the fuselage under water, and the wings on the surface. This was definitely not a good platform from which to ditch in the water. The navigator sat in the web seats aft of the first two rows, in the cargo compartment, for safety. The more I reflect on it, the aircraft was a piece of crap.

Special Operations Mission
The mission of the unit was special operations, which consisted of low-level flying over the Berkshire Mountains and ending with a low-level parachute drop of cargo at the designated drop zone at Westover AFB. On late afternoon summer days, flying made for a queasy stomach, what with thermal updrafts and downdrafts, at 1500 ft. above ground level, bouncing around and breathing the air with the residual smell from the chemical in the aircraft. Low-level navigation in this aircraft consisted of map reading, while seated in a sling in the cockpit entry door, looking for some small turning point, with navigation legs of 10 to 12 minutes. When I couldn't see the turning point, I turned on the elapsed time, and looked for the next point.

For instance, I remember looking for two small ponds just off the intersection of two small roads. I saw two small roads, but was not sure they were the right roads, and there were no small ponds. It turns out the ponds dried up in the summer. How was I to know?

We were flying at about 110 knots airspeed (approximately two miles per minute). After the low level navigation exercise, the procedure was to approach the drop zone low and kill the aircraft drift, as observed by a smoke flare from the drop zone personnel. I called for "Green Light" over interphone at the CARP (Computed Air Release Point), while the copilot turned on the green light switch and the light at the loadmaster station in the rear of the aircraft turned from red to green.

The pilot then pulled the aircraft nose up into a steep climb, pushed the throttles to maximum power, the copilot toggled the two jet engines to 100% speed (power) and the loadmaster released the tie downs for the load. Meanwhile, my stomach dropped to my toes as the aircraft abruptly pulled up, and the load rolled off the rollers on the cargo floor out the back of the aircraft. This ceased to be fun in a hurry.

I was never comfortable with map reading at low level, and didn't develop this skill in the short time I was associated with the C-123K. Flying at night was even worse, because all I had to work with were the lights from small towns in the Berkshire Mountains. I tried to gain the assistance of the pilots to help me find the miniscule ground checkpoints, using the argument we were all in this together, and if we busted the drop zone, we would all be in trouble. This worked for a brief time in flight, but then the pilots degenerated into *yak yakking* and I was once again on my own.

I fully knew if this were a combat situation, the pilots would work with me, as our lives would be at stake, but in the Air Force Reserve, it didn't work that way. The navigator sat in a sling, which resembled a short hammock, which he slung across the entry to the cockpit, after takeoff. He had the chart, with the route plotted, attached to a clipboard, and basically used map reading for navigation.

The Air Force had bad experience with the C-123 type aircraft while resupplying Khe Sanh in the northern section of South Vietnam. The Marines were in a constant siege by the North Vietnamese, and every aircraft approaching to unload supplies got the treatment. The C-123 was so slow it got shot up pretty badly as it approached the runway on the hilltop. This aircraft was withdrawn as a resupply aircraft in favor of the faster C-130 aircraft. In an active duty call up, I could see our unit used in special operations missions, most likely at night.

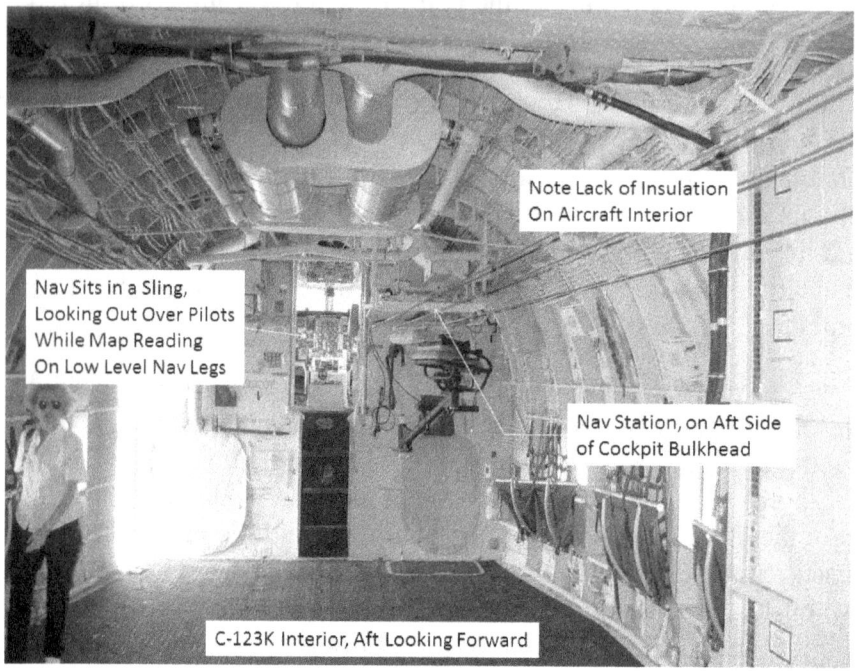

Note Lack of Insulation On Aircraft Interior

Nav Sits in a Sling, Looking Out Over Pilots While Map Reading On Low Level Nav Legs

Nav Station, on Aft Side of Cockpit Bulkhead

C-123K Interior, Aft Looking Forward

Interior of the C-123K Aircraft. Note the lack of sound suppression inside the fuselage. Photo by Author

C-123K Aircraft History (Ref: Wikipedia)

"The XCG-20, also known as the XG-20 was the designation applied to a large **assault glider** developed immediately after **World War II** by the **Chase Aircraft Company** for the Air Force, and was the largest glider ever built in the United States. The XG-20 did not see production due to a change in Air Force requirements; however, it was modified into the successful **Fairchild Aircraft C-123 Provider**. This was powered by twin Pratt & Whitney R2800 engines (2300 Horsepower each) driving Hamilton Standard three-bladed constant speed, full feathering and reversing propellers. This Short Takeoff and Landing (STOL) transport aircraft saw extensive service in the

Vietnam War. The C-123'K' version was retrofitted with twin GE J85-GE-17 booster turbojets of 2850 pounds static thrust each. These jet engines increased the aircraft payload by a third, shortened its takeoff distance, improved its climb rate, and gave a much greater margin of safety should one of the piston engines fail. It didn't do much for the aircraft top speed, however."

C-123 Provider
Ref: commons.wikimedia.org

The C-123K saw extensive service during the Southeast Asia War, both as a "trash hauler" and as an Agent Orange defoliant spray aircraft. This latter operation was called Ranch Hand. In order for the chemical to be effective, the aircraft flew low and slow. It made a tempting target for the bad guys on the ground, and the aircraft were frequently shot up.

C-123K Provider
Ref: commons.wikipedia.org

UC-123B spraying during Operation Pink Rose 1967
Ref: commons.wikimedia.org

Why do I go into such detail about Operation Ranch Hand? Some of the C-123K aircraft assigned to the 731st TAS were former Ranch Hand aircraft. Leaking Agent Orange had permeated the cargo floor and could not be removed. There was a permanent smell while inside the aircraft. It did not have much life left, and the Air Force was not going to put any money into them. I pray the chemical will not have any effect on me as I age.

The Veterans Administration contacted me about inclusion in their Agent Orange Database. I agreed and underwent a physical examination at the VA facility in Gainesville, Florida. Eligibility for the database was based on setting foot on the ground in areas of Southeast Asia susceptible to exposure to the chemical. I had volunteered for a C-124 long range cargo mission to Cam Ranh Bay in Vietnam in 1969. I was on the ground for about two hours while an aircraft fault was cleared, and while cargo was offloaded and uploaded. I volunteered information I flew as a navigator on former Ranch Hand C-123K aircraft in the AF Reserve, and this was a more likely exposure. The VA wasn't interested. It didn't fit the narrative. Government insanity!

Aircraft Performance
Just a note of clarification; the specifications on the aircraft quote a maximum speed of 240 mph, range of 1825 miles and ceiling of 28,000 feet. The way aircraft performance is quoted, each value stands alone, not in conjunction with the others. In this case, the aircraft may have attained a top speed of 240 mph, but not at the same time attain a range of 1825 miles at 28,000 feet altitude. These specs may seem impressive, but personal experience provides much more subdued values. The aircraft cruised at 100-110 knots true airspeed and was unpressurized, so we flew under 10,000 feet altitude without oxygen.

Airdrop Exercise at Pope AFB
I volunteered to fly in a large airdrop exercise at Fort Bragg (Pope AFB), North Carolina. We flew in six of our C-123K aircraft down to Pope AFB from Westover AFB, Massachusetts for this exercise.

Engineer, Set Refueling Power

There were airdrop aircraft from active duty, Air Force Reserve, and Air National Guard units from all over the country. Most aircraft were C-130 aircraft. There were a few De Havilland Buffalo and Caribou aircraft from Army aviation units.

It took six hours to fly from Westover Air Reserve Base to Pope AFB. By the time we landed, my hearing was impacted. It recovered the next day for the exercise, but I had the same problem on the return flight. I was definitely developing a hearing problem associated with loud noise exposure in aircraft.

De Havilland Buffalo Short Takeoff and Landing aircraft This aircraft was similar to the De Havilland Caribou, except for the GE T-64 Turboprop engines. Ref: Wikimedia.commons.org

Engineer, Set Refueling Power

De Havilland DHC-4 Caribou with piston engines   Note this piston powered aircraft was
assigned to the U.S. Army
Ref: Wikimedia.commons.org

Our C-123K aircraft were <u>completely</u> worn out at this point in time, even more than the C-124 aircraft I previously was associated with. The engines were beyond their mandatory interval for overhaul, and were operating on official waivers. It was common to see engines pumping out blue smoke from worn piston rings, just like in old worn-out cars. The Air Force just wasn't going to put any money into maintaining these aircraft. Overhauled engines were not available so the existing engines stayed on wing with waivers.

The airdrop exercise consisted of flying a low-level navigation route and dropping a small parachute load over the base drop-zone. Since I was fairly new, and had not flown this low level route before, I was nervous. We were in the second aircraft in the formation. While warming up the engines before takeoff, the lead aircraft pilot called us on UHF radio, "Number two, this is Lead. We are having a mechanical problem. Be prepared to assume Lead responsibilities." I swallowed hard. I observed blue smoke pouring out the lead aircraft's engine exhaust. Just before takeoff, Lead called us and told us he resolved the problem and resumed lead responsibility. I had a sigh of relief. The mission came off without a hitch.

Potential for Recall

I recently read a memoir by an Air Force pilot who flew the C-123K in Laos during the Vietnam War. The book was titled, *Flying Past Midnight*, by John T. Halliday. The aircraft flew four-hour flare-drop missions at night. There were two shifts, each aircraft flying four hours in turn. The parachute-supported flares kept the battlefield illuminated in front of the friendly Laotian forces to preclude the bad guys from sneaking up on them at night. The aircraft also flew reconnaissance missions over the Ho Chi Minh Trail, at night, looking for heavy equipment. The C-123K were cheap to fly and had good endurance. Now I understand why we were kept as a unit. Ultimately, the aircraft were retired.

Government Insanity

LATE FLASH!

For years, crew members who had flown the C-123K in the Air Force Reserve pressed the government to include them in the Agent Orange Disability category. The government fought these retired crew members tooth and nail. Finally, the VA caved and has specifically identified these crew members as eligible for a VA Disability. Since Type 2 Diabetes is a symptom of exposure to Agent Orange, proving exposure was no longer a requirement.

As of June 2015, the VA has identified those aircrew members who flew the C-123 in the reserves, after the Vietnam conflict, as being in a special category of interest. I immediately filed a disability claim with the VA, as my Type 2 Diabetes is a symptom of exposure to Agent Orange. I was told my claim processing time could be as long as six months to a year or longer. I suppose the government is hoping we all will die before the claims need to be acted upon.

Shortly after filing my claim, the VA formally disapproved my disability, based on my inability to prove my diabetes was caused by direct exposure in the brief time my C-124 aircraft was on the ground in Vietnam. The VA completely missed the fact I flew as a crew navigator in C-123K aircraft that had previously been spray aircraft for Agent Orange. The VA ruling specifically identified the 731st TAS as evidence of exposure to Agent Orange. I immediately refiled a claim with copies of my orders assigning me to the 731st TAS at Westover AFB. Anyone wanting the health care system in the country to go single payer should deal with the VA.

The Federal Government does nothing efficiently, with one exception: the military. The reason is the military mission is to break things and kill people and they are darn good at it!

GE J85 Booster Engines

C-123K Tactical Airlift Transport
Air Force Photo

I tried to catch up with my required flights

As a reservist aircrew member, I was authorized 32 annual Additional Flying Training Periods (AFTPs) for which I was paid base pay and flight pay (About $32/day net in then-year dollars). The 731st TAS made our C-123K aircraft available Tuesday and Thursday afternoon and evenings. I was strongly encouraged to come out to Westover Air Reserve Base during those days in order to accomplish my annual navigator's requirements for day and, especially, night airdrops. Since I worked about 100 miles away, it was difficult for me to commute to Westover AFB. I decided to cut short a Tuesday at work, drive to Westover in time for the early afternoon mission, stay for a quick bite for supper, and fly the evening mission. After the two flights, I had trouble staying awake on the two-hour drive home. I was as tired as I can remember. The next morning, I had trouble getting out of bed to go to work. I refused to do this again.

C-123K Twin Engine Failure

I signed up for a cross-country flight. The aircraft departed Westover Air Reserve Base on Friday evening, and about 45 minutes later, in the Albany, New York area, one piston engine blew a hole in a cylinder head (known as a jug). The pilot shut the engine down, but as soon as the aircraft was trimmed for single engine operation, the other piston engine torque went to zero, and climbed ever so slowly. The pilot restarted the first engine and shut down the second. The copilot quickly restarted the two jets (which were routinely shutdown after takeoff). We returned to Westover and made an emergency night landing with the fire trucks following us down the runway.

As soon as the aircraft speed slowed to the point where the leaking fuel from the running bad engine was no longer washed away from the hot nacelle, a fire ignited. I saw the orange flames reflected inside the fuselage as the fire trucks caught up with us and put out the fire. That worn-out aircraft was a disaster waiting to happen. It was absolutely shot!

"Summer Camp" was Accomplished at Howard Air Base, Panama

Our annual active duty was performed at Howard Air Base in Panama. There were C-123K (the aircraft belonged to Panama) aircraft assigned to the base. Our aircrews were flown down to Panama by a C-130 aircraft. We then flew local missions spraying the jungle for mosquito control. I signed up to go on one of these active duty tours, but left the squadron before I actually was deployed.

Hoping for More Modern Aircraft

I held on as long as I could, hoping the Air Force planned to replace these aging aircraft with something more modern. This was the timeframe the KC-135 jet tankers were being moved into the Air Force Reserve and Air National Guard, and we anticipated having the KC-135 assigned to Westover, with its long runway. One of our pilots was promoted to brigadier general, and assigned to AF Reserve duty in the Pentagon. He visited us one training weekend and was besieged with inquiries as to when we were going to get a more modern aircraft. He couldn't tell us anything, as that information was not for public dissemination. He did encourage us to hold on and hinted something was in the works. This added fuel to the fire sustaining the rumor we were going to get the KC-135 jet powered tanker. It was not to be.

I moved on as I lost patience, and my flying career as a navigator came to a permanent end. My decision turned out to be a good one. Instead of replacing the C-123K with a more modern aircraft at Westover, the unit was moved to Peterson Field, Colorado, and changed into a C-130 airlifter unit. It would have been the end of my flying career anyway. As a matter of interest, the Ohio Air National Guard unit I left when I transferred to Massachusetts was the first unit to have the KC-135 jet tanker assigned. If I had stayed in Ohio, I would have extended my flying career, and would have jet tanker experience to boot.

Reflections on my Flying Career

When I left active duty in the Air Force, I worked as a propeller installation engineer for four years. Every aircraft I was associated with as a navigator was powered by propellers. Watching the slow demise of the piston engine-powered fleet of aircraft convinces me I made the correct decision to leave the aircraft propeller engineering field.

Every aircraft I was associated with was retired and chopped up for scrap. This saddens me. I feel old! This includes the T-29 Flying Classroom, KC-97 Stratotanker, C-124C Globemaster II and C-123K

Provider.

Overall Assessment of Career as a Navigator

Overall, the experiences I had over the years flying as a navigator were the best of my life, professionally. I met a great group of people and traveled to a lot of exotic and not so exotic places. I had a tremendous amount of responsibility as a navigator directing an aircraft either to a refueling or delivering cargo, both long haul and short field. Although, in my civilian profession as a mechanical engineer, I had a lot of responsibility, never did I have as much as in the military. Today, with the technology advances of GPS navigation systems, navigators are no longer used. The old KC-135 tankers are still flying on active duty and in the reserve forces. They have been retrofitted with systems making the navigator obsolete.

I am grateful I lived in an age where I had the opportunity to train and perform as a navigator. I wouldn't change a day in my life in the military, and would do it all over again, given the opportunity.

I have to make the following observation, based on having been assigned to two flying units in the Air Force Reserve and one in the Air National Guard. While in the Reserve, I had the feeling you had to look out for yourself constantly, whereas in the Air National Guard, the organization seemed to look out for you. Let me give examples.

After every flight, say an AFTP, paper work had to be filled-out in order to be paid. The paperwork was placed in a mail "In" box. It was possible you would not be back to the base for a week or more, and if the paperwork was made out wrong, it was bounced back and placed in your mailbox for correction. Meanwhile you would not be paid. When returning from a long trip, on a 15-day active duty tour, you would be fatigued, grubby and sweaty. Again, the paperwork had to be exact. Your pay depended on it.

In contrast, with the ANG, mistakes in the paperwork were corrected by the administrative clerk, and submitted the next duty day. Pay was always prompt. On Operation "Creek Party" deployments (15 day active duty tours), the unit actually submitted the paperwork for you, even before you returned from the deployment, and the check sometimes beat you home! I would take the Air National Guard over the Air Force Reserve anytime!

I read in the Hanscom AFB newspaper there was a new Hanscom-based Air Force Reserve unit being formed. It was associated with the Aeronautical Systems Division (ASD) at Wright Patterson AFB, Ohio. This unit, actually a detachment, was to consist of procurement officers who would be slated to move into ASD offices at Wright Patterson AFB, when the *Balloon* went up, to occupy slots opened up by virtue of active duty rated officers (pilots and navigators) being recalled back to the cockpit. ASD was the organization procuring major aircraft weapon systems (aircraft and engines), issuing contracts, monitoring contract compliance, and supporting the weapon systems in service. The difference in the unit compared to a flying unit is apparent in the following comparison, relative to participation and pay (Category A is a flying billet whereas Category B is non-flying.)

	Category A (Flying)	Category B (Non Flying)
Weekend Drill	Every Month	Every Other Month
	(4 Days Base Pay/Flight Pay)	(4 Days Base Pay)
Add'l Flying Training Period 32/Year		None
	(Base Plus Flight Pay)	
Annual Active Duty Tour	15 Days	15 Days
	(Base, Flight, Quarters, Subsistance Pay)	(Base, Quarters, Subsistence Pay)

End of Flying Career

This new unit looked like a glove fit for my background as a senior engineer at General Electric Small Aircraft Engine Division. I immediately expressed interest and was picked up by the commander of the Boston Detachment.

The position had such promise, and I was excited, both to get out of flying in the C-123K, which I absolutely hated, and to be productive as an engineer. Alas, it was not as I had expected. I quickly found out the engine system project offices at Wright Patterson would not let me walk in their office area, because there was a perceived conflict of interest. In other words, if I worked on a project concerning P&W engines, our competitor, Pratt and Whitney could complain I was being exposed to their proprietary information.

And GE could complain I was short-circuiting any initiatives the company was working on with the Air Force. Whoever said there were no political aspects in engineering? As a result, I had to work on projects on which I had no knowledge or background. Air Force insanity continues.

I put up with this, as I needed at least another five years to get my 20 years for retirement in the Air Force. A good example of the atmosphere is as follows: I put together a group of reservists in my detachment to work on a project submitted by a civil servant at Wright Patterson AFB. We were to determine why officers did not want to be assigned to system project offices on older fielded aircraft or engines. The civil servant was sincere, but he wanted us to talk to his boss before we started. The boss had a fairly high level position, with a large impressive office. He made it abundantly clear he would go along with us working on this project, but he did <u>not</u> want us to come up with any conclusions making waves!

We got the message loud and clear. We spent the next year, working part-time on weekend drills every two months, and came up with a report. Our sponsor, the civil servant, was irate over our conclusion, and actually gave us an unsatisfactory feedback rating. This rating reverberated up and down the organization, with lots of mashed teeth. I explained it away by citing the high level boss's initial instruction. The reader must understand in an organization like the Air Force, everybody is outstanding, and everything is excellent. That is the civil service way. So, when an unsatisfactory report is submitted, you can guess how it was received.

Anyway, I hung in with this ASD organization, way too long. I got my five needed years and hung in there for an additional four years. Looking back, I can only explain it by virtue of wanting to be paid. I came to dislike the ASD position almost as much as the dislike for the C-123K aircraft.

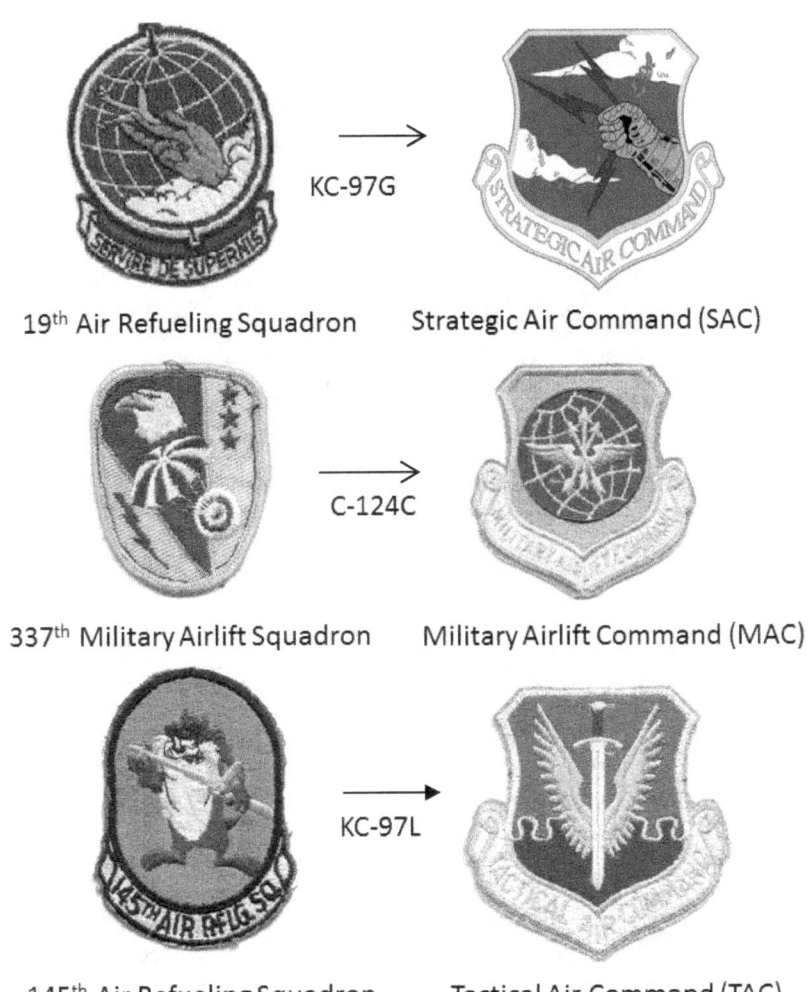

19th Air Refueling Squadron   Strategic Air Command (SAC)

337th Military Airlift Squadron   Military Airlift Command (MAC)

145th Air Refueling Squadron   Tactical Air Command (TAC)

Unit Patches
Art Provided by Author

Chapter 12: LIFE AFTER WORK

Interest Kindled for Antique Cars at Interpersonal Awareness Seminar

1948 Chrysler Similar to Car Owned by a Peer at GE
Ref: Wikipedia.org

In 1979, I attended a GE-sponsored, offsite, weeklong seminar entitled, "Interpersonal Awareness." This experience changed my life. One of my classmates was a technical writer at GE, and he owned an antique car, a 1948 Chrysler Sedan. He brought the car to class one day and I was hooked.

One exercise in particular stuck with me. We each had a journal in which we entered any information we wanted. The moderator had us select a new page, and draw a line across the top of the page. Then we put tick marks, starting from the left, with our date of birth, today's date in the middle, and on the right, our day of death. During this time, there was mood music playing from a tape recorder. Wow, this was getting heavy. Under the line we listed all the things we wanted to do in our life we had not accomplished.

Today, we would call this our bucket list. Next to each item we wrote what was preventing us from accomplishing the desired item. I suddenly realized I wanted to restore an antique car. This exercise plus exposure to the 1948 Chrysler brought this desire from my subconscious into my conscious.

One of the other attendees had an intelligent boss, but who managed as a tyrant. GE was full of managers like him. So was Hamilton Standard. This individual brought up the subject of how to deal with his boss. The moderator told him he could "Manage Up." What a stupid idea.

The Search for a Project Car

My two sons were in their senior year in high school, and college was coming up fast. I never did anything for myself, and I decided I was going to restore an antique car, in spite of the looming college expenses. I started searching the antique car section of the local *Want Ad* book. I was particularly interested in a Ford or Mercury of 1950s vintage, because I remembered the melodious exhaust sound, with dual exhaust and glass pack mufflers.

There was a large auto body shop across the street from GE, where I worked. On the body shop front lot were two cars for sale I had an interest in. One was a beautiful restored 1950 Ford Convertible painted fire engine red with a white top. I passed because I wanted to restore a car rather than buy one already restored. This car was one of those that got away. It was a nice car. I should have bought it.

Also, on the lot was a 1950 Mercury convertible in sorry shape, needing complete restoration. The car was on consignment and I looked up the owner. I took a test ride in the car, but got scared about the degree of work it needed. It had a price of $3000, and I should have snapped it up. This car would be worth about $90,000 today in fully restored condition. Again, I passed. I saw a 1951 Mercury Convertible, in sorry shape, in front of a local strip club on U.S. Route 1 in Peabody, where I lived. I never stopped to look at the car as it was incomplete and needed total restoration. I was such a fool, as the car I eventually bought needed much more work than this one did.

1950 Ford Convertible
Photo by Author

I heard from the grapevine there was a 1940 LaSalle in the back of a service station in Chelsea, Massachusetts. I drove over to look at it. The car was parked in the rear, had been in a collision, and needed restoration. Again, I passed, but I should have bought it. I believe it had a price tag of $6000.

The Find

A friend told me about a contractor in Lynn who had a 1939 Ford Convertible Coupe for sale. He had disassembled it and stored the components in two rented garages. I arranged to see it.

I fell in love with a picture postcard of what the car looked like restored. It didn't hurt the year of the car was the year of my birth. I purchased the "*car*," which was really a collection of rotted out parts, for $1500.

The owner had purchased this "*car*" on Cape Cod and had disassembled it using a torch, as necessary. The *car* needed a total restoration. I was foolish to undertake this level of work, as I could have had a drivable antique car for much less money than I would need to spend to bring this one back to original condition. I guess I was in love with love, as they say.

Postcard of 1939 Ford Convertible Coupe with Rumble Seat
Photo Provided by Author

There was a twist to the car's history. Before World War II, Ford had an assembly plant in Copenhagen, Denmark where Ford produced cars for the Scandinavian market. Ford of America shipped kits of parts to its plant in Denmark, and the cars were assembled and upholstered there. This car had been assembled in the Denmark plant. It had trafficators for directional signals as original equipment (see photo), an export sheet metal tag on the driver's side floor, and a European data plate on the engine firewall. I am sure this *"car"* survived World War II in Europe. Boy, what stories it could tell if it could talk! I paid $1,500 for the collection of distressed parts.

Locating a Restoration Shop and Securing Financing
After my two sons spent the morning taking their SATs for college entrance, we all went to pick up the parts. The owner had a dump truck, and he offered to help deliver the *"car."* As we drove by neighbors on the rear facing street from our house, I recall one saying, "He'll never finish that."

I was determined. I spent the next six years pecking away at the restoration as I had a dollar or two available. I thought the hard part would be finding a source of financing. Suddenly, a new loan product became available, called a home equity loan. I secured the necessary financing, and set about to locate a restorer. This, it turned out, became the hardest problem. I was turned down by two restorers as they were afraid of the immensity of the task.

Finally, I located an individual who was willing to take on the project as a challenge. He had the skills, as he was a precision sheet metal mechanic, by day, and an antique car restorer by night. He could do anything with sheet metal. He estimated the job would require about $10,000 on my part, based on his $15/hour labor rate. Ultimately, he was off by a factor of two. For $20,000, I could have purchased a 1939 Ford Convertible Coupe from California, without ever having a rust issue.

My 1939 Ford Convertible Coupe Deluxe Restoration

The Condition of the *Car* as Restoration Started
The engine had a crack in the cast iron block and was of 1938 design, not 1939. Up until 1938 each engine head had 21 studs and nuts holding it to the block. Thereafter, a design change was made to have 24 studs per head. The transmission and rear end appeared to be OK, but there is more to this story later. I was missing one rear fender, but had three front fenders, and I was missing the speedometer. There was no heater or radio, but this could have been the way the car was configured, as a heater or radio was an option in those days.

The rumble seat bottom cushion was missing and in its place were the rotted remains of two jump seats facing each other. The convertible top bows were intact. This was what I had to start with.

The Engine
I needed to make a decision on what to do about the engine as it was about one year older than the car. I found a 1941 Ford parts car on a used car dealer lot in Lynn, Massachusetts, where I worked. I paid $350 plus $25 for a tow to my house. The 1941 Ford had the correct appearing engine, with 24 studs per head. I started stripping the Ford

for parts I could use in my restoration and parts I could sell, as they did not fit my car. The transmission was a column shift as opposed to the floor shift on the 1939 car. I sold the transmission for $50 at an antique car flea market.

1939 Ford Cabriolet
A Ford Year in Transition

1939 Ford Models	*1940 Ford Models*
Floor Shift ⟶	Column Shift
Rumble Seat ⟶	Gone
4-Door Convertible ⟶	Gone
Take Apart Head Lights ⟶	Sealed Beam Headlights
Direct Reading Engine Temp Gage ⟶	Ignition Key Operated
Manual Convertible Top ⟶	Vacuum Powered Top

1939 Ford Offered for the first time

Hot Water Heater (Previously Hot Air Heater)
Hydraulic Brakes (Previously Mechanical Brakes)

1939 Ford Cabriolet

Assembled and Upholstered in Ford Copenhagen Plant from Kit shipped from USA

Flathead V-8 Engine, 85 hp
Transmission 3 Speed, Floor Shift, 2nd/3rd Gear Synchromesh
6 Volt Electrical System, Positive Ground
Hot Water Heater
Rumble Seat
Rotomatic Radio
Directional Signal "Flippers" (European)

I Opened the Transmission. "Yikes!"
Later, when I opened up the transmission for the 1939 Ford, instead of heavy oil inside, I found about one inch of rust colored water. The oil had been drained and water had seeped in. The gears were corroded and were worthless. I later found out the gears, shafts and bearings from the column shift 1941 Ford transmission were interchangeable with the floor shift model.

I screwed myself by selling the column shift transmission. Now I had to find a floor shift transmission for my car. The lesson I learned was never to sell any parts until the restoration was completed.

Ultimately, I made the $375 back I invested in the 1941 Ford parts car, and I now had the engine I needed. I found a good 1936 Ford floor shift transmission at an antique car flea market and bought it for $65. This transmission was interchangeable with my 1939 Ford.

My restoration shop owner picked up the collection of parts in his pickup truck for transport to his garage. Initially he transported the sheet metal parts to an Easy Strip location to have the paint and rust removed, chemically. Then the restoration started in earnest.

I Bought an Engine, Transmission and Rear End from the Dimmest Bulb in the Chandelier
I found an ad in the local Want Advertiser, for a 1940 Ford engine. I called the seller and found out he had replaced the clutch in his 1940 Ford years ago, and the clutch still slipped. He then stripped out the engine, transmission and rear end and junked the rest of the car. The engine, transmission and rear end were stored outside behind his house with a tarpaulin over the parts. The seller was located on the other side of Peabody, Massachusetts.

I offered this person $50 for the engine. He didn't respond, so I told him to think about it. The next day, I called and increased my offer to $100 for the engine, transmission and rear end. This individual definitely did not have both oars in the water. He took my offer. My father-in-law had given his old 1973 Ford station wagon to my son, so we hauled the parts home in this car. The purchased engine was seized, but this was no surprise, and the hole in the intake manifold where the carburetor bolted on had been the home of what appeared to be mice. I decided to use this engine instead of the one from the 1941 parts car.

When I took the heads off this engine, and exposed the engine serial number, it turned out to be a 1939 engine rather than a 1940, so it worked out. The main difference was the 1939 engine had a rear "slinger" or windback for the rear bearing oil seal as opposed to an actual rear seal. The consequence was when the car was parked on a hill, nose up, without the engine running, there would be a small oil leak out the rear slinger. I examined the clutch and found oil had permeated the facings, so this explained why the clutch slipped when the former owner replaced it.

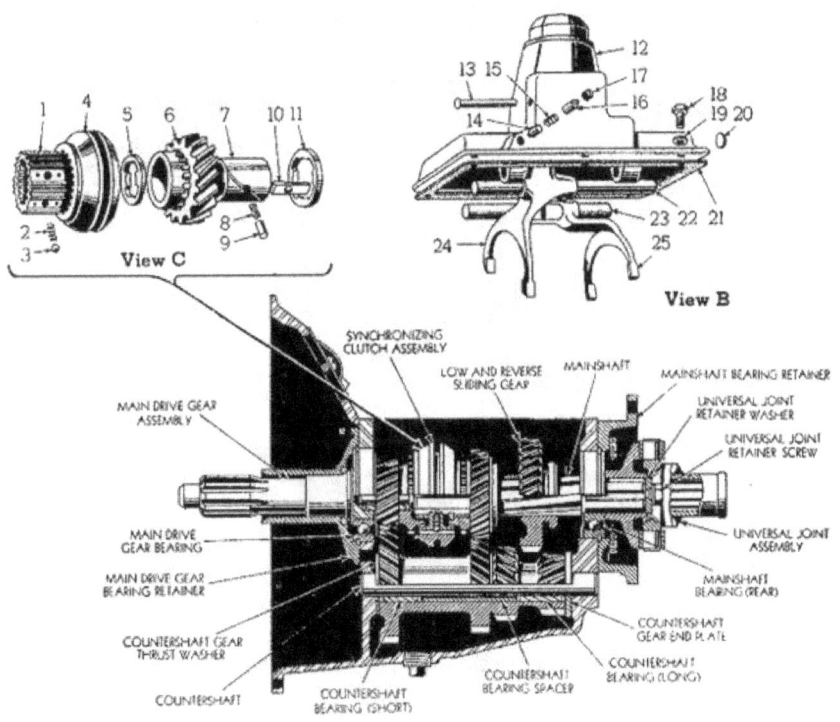

Ford Top Shifter Transmission
Ref: Ford 1939 and 1940 Engine & Chassis Repair Manual

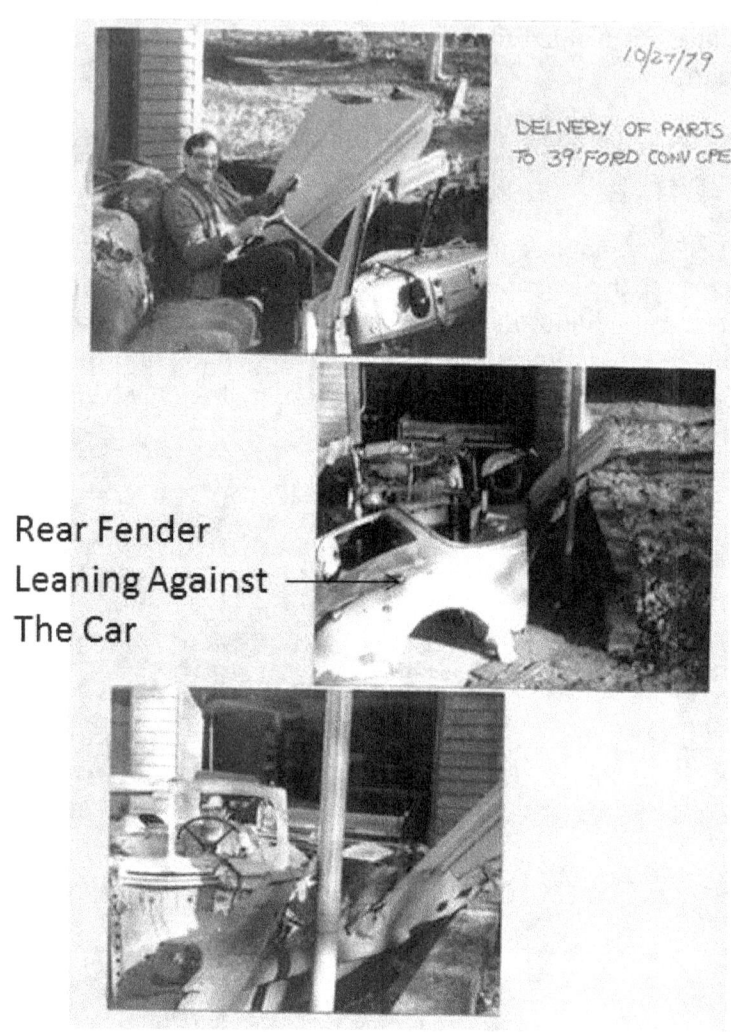

10/27/79

DELIVERY OF PARTS
TO 39' FORD CONV CPE

Rear Fender
Leaning Against
The Car

10/27/1979

Delivery Day for the Collection of Rotted Parts Comprising a 1939 Ford Convertible Coupe
Photo by Author

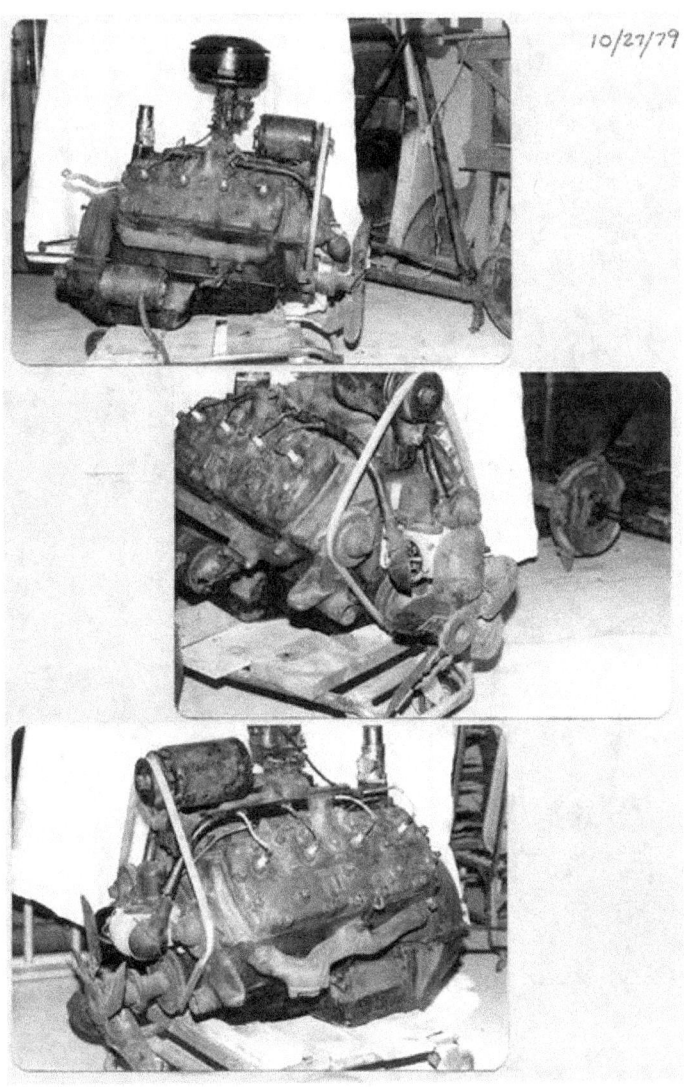

Ford Flathead V-8 Engine as Delivered to Author in 1979
Photo by Author

Restored 1939 Ford Convertible Coupe
Owned by Author

European Trafficator
Both Photos by Author

Note the Wood Grained Dash. The Wood Graining is Actually Paint, as this is a Ford
Photo by Author

Ford Flathead Engine
Ref: en.wikipedia.org

The "Hulk" on the way to the chemical stripper for rust and paint removal
Photo Provided by Author

I Bought a Frame from a 1938 Parts Car

The frame was not salvageable. I found a 1938 Ford Sedan being broken up for parts, in the local Want Advertiser. I wanted the frame but the car was still assembled. The owner agreed to remove the body from the frame over the next week. He also sold me the "banjo" steering wheel and steering shaft assembly, which was correct for my car. The steering wheel on my car was the configuration for the 1939 Ford Standard, as opposed to the Deluxe.

Once again the 1973 Ford Station Wagon came in handy. We loaded the rusty frame from the 1938 Ford into the station wagon and took it to a small foundry in Lynn. They sandblasted it for $75. I then set up the frame in one bay of my two car garage, on jack stands, and proceeded to paint it with spray cans of primer and then multiple coats of black matte finish paint.

Next I started assembly of the brake system, consisting of the brake lines or steel tubing, brake hoses, master and slave cylinders, the front and rear springs, and so forth.

Engine Overhaul

I contracted with an engine rebuild shop, near where I worked, to do the engine machining work. The engine was indeed seized. The pistons were removed using a sledge hammer and chisel to completely destroy them. Likewise, the valves were demolished. The engine block was then "hot tanked" to remove the entire residue from the coked oil. Then the shop accomplished the machining of the cylinder bores and valve seats. It cost approximately $800 for all the machining, which would be a bargain today. I brought back the fully machined engine block to my house so I could build the engine back up with new fitted pistons, new valves, camshaft, machined crankshaft and oil pump.

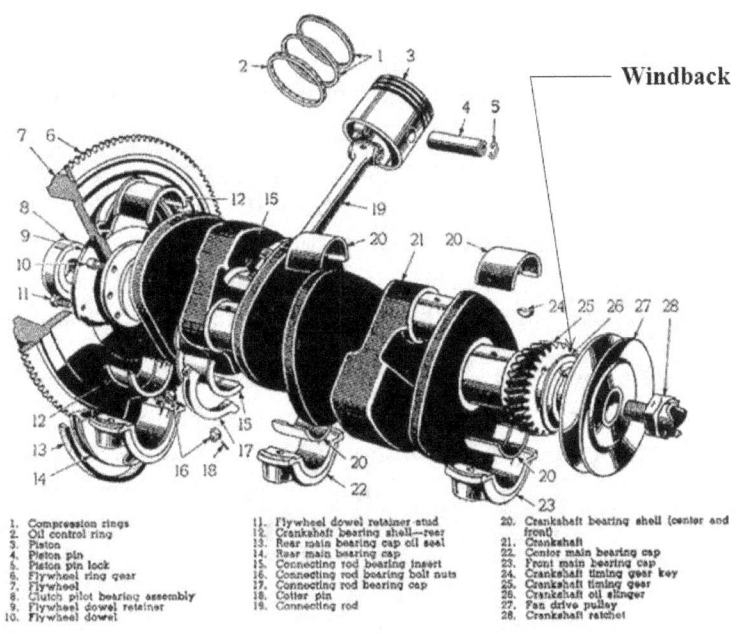

Ref: Ford 1939 and 1940 Engine & Chassis Repair Manual
The windback shown as the front seal on this model was used also instead of the rear seal depicted.

353

As I rotated the machined block on my engine stand to get a good angle for reassembly, the light caught one cylinder bore just right, and I saw what I thought was a fine crack in the machined bore. I brought the block back to the shop, and they confirmed the crack. The shop then bored the cylinder oversize, pressed a sleeve in the bore and remachined the bore so the piston would fit properly. I hate to think what would have happened if I had not spotted the crack and completed the engine buildup. Since the shop charged me for the sleeve and subsequent machining, I felt they did not treat me very well, particularly since the shop should have spotted the crack when they did the original machining.

I proceeded to build up the engine with new or rebuilt parts. Upon completion of the buildup, I painted the engine with authentic Ford Green paint, and put it aside.

Business Agreement

My restoration shop owner suggested we make an agreement to our mutual advantage. When he needed to fabricate a sheet metal part not available on the restoration parts market, he would develop the tooling and make the part for my restoration. He would not charge me for the time it took to develop the tooling, but charge only for the time it took to actually make the parts. He would then be free to use the tooling to make parts for the restoration market.

Many parts on my 1939 Ford fit a variety of other year Fords. For example, the lower door sections were not available, but were shot on my car. He developed the tooling to make the lower door parts, and the parts fit 1935 through 1940 Fords. He made lower door parts as well as many other parts for Fords undergoing restoration all over the country and the world.

Initial Engine Startup

It was time to start the restored engine for the first time, in my garage. The body was still at the restoration shop. I added gas to the restored gas tank, strapped a six volt battery to the chassis, and cranked the engine. I could not get the engine to fire. I was frustrated, and called a friend, from work, to give me a hand.

He quickly determined no fuel was being fed from the gas tank. I had previously poured "sloshing fluid" inside the empty tank, to seal all the pin hole leaks and capture any rust particles. After sloshing this liquid around to insure all the surfaces in the tank were wetted, I emptied the remainder. What I did not know was the compound not only blocked any pin holes, it also blocked the tube in the tank feeding the gas into the gas line on the way to the engine. After unblocking the feed tube in the tank, the engine fired off. What a feeling it was, as the engine fired off and ran for the first time!

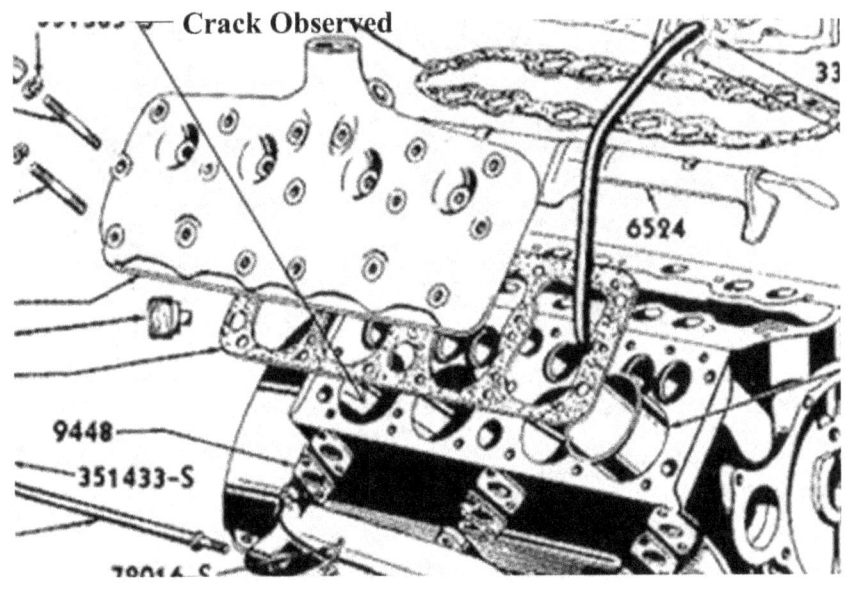

Ref: FORD CHASSIS PARTS and ACCESSORIES Catalogue
Passenger Cars 1928 thru 1948, November 1950

Chassis Assembly

The chassis of a car is basically what is left when the sheet metal body is removed. It consists of a frame, suspension, brake system, and power train (engine, clutch, transmission, drive shaft, and rear end). I installed all these items on my sand blasted frame after I painted it. Wow, this was now starting to look like a car! My restoration shop had my "bad" frame and used it as a fixture to build up the sheet metal body as it was restored. The plan was to lift the completed sheet metal body off the "bad" frame, slide the nicely painted restored chassis under the body and lower it onto the chassis. This was what he did.

The body is mocked up on the original frame, which was used as a fixture. I actually located and used another "good" frame off a 1938 Ford Sedan parts car.

Photo provided by Author

Cost Inched Up

The shop started the restoration charging $15/hour labor rate. This was very fair, as the owner multiplied the hours expended by an efficiency factor. So, if he spent two hours on my job, but had to answer the telephone and talk to a customer for a few minutes, he multiplied the two hours by 80%, thereby charging me 1.6 hours. As time went by, the shop needed to increase its hourly rate, as the owner wasn't making any money. By the time the task was completed, his rate was $22.50/hour, which was still fair.

Trip Home from Restoration Shop

After two years in the restoration shop, the time came to drive the car home to Peabody, Massachusetts. The route required a leg on the highway of approximately 25 miles as well as back roads at the beginning and end of the journey. This was the maiden voyage for the completed car, and I was nervous. Initially, everything went well, but as the miles added up, the car slowed perceptibly, and the engine temperature gage showed hotter than normal. This situation got progressively worse, until I had to pull over to investigate.

I noticed my brake taillights were illuminated constantly. I let things cool down and moved out again, with the same results. After many stops, I made it home. Investigation revealed the hydraulic brakes were not adjusted correctly.

There was a bypass port in the brake master cylinder. This needed to be uncovered by the shaft from the brake pedal, to allow brake fluid to return from the wheel cylinders as the fluid got hot, with normal driving. When this bypass port was not open, as the fluid heated up, it expanded and basically applied the brakes. This was why the brakes came on by themselves, the brake lights lit and the engine got hot trying to overcome the drag of the brakes.

On the right is what was left of the steel floor of the rumble seat. On the left is a new floor fabricated by the shop. Photo Provided by Author

The completed car was loaded on a trailer for transport to the upholstery shop. The car was upholstered there and the convertible top installed. Photo Provided by Author

Ford Flathead Engine Inherent Defect

Once I started driving the car, I learned about the inherent defect in the design of the Ford flathead engine. The exhaust gases passed through the engine block before they exited the engine. By doing so, the exhaust gases added an additional heat load to the engine cooling system not existing in other engine designs. The Ford flathead had a reputation in its day of running hot. This resulted in a common malady, called vapor lock. The fuel pump was located on top of the engine, right where the heat load was high. At idle speed, when the engine was hot, the fuel went into its vapor state rather than liquid state. The fuel pump could not "suck" vapor fuel, only liquid fuel. The engine then simply stalled, as it ran out of fuel in the carburetor.

Is it Vapor Lock or an Overheated Coil?

There is a school of thought believing vapor lock wasn't the problem; rather the coil insulation between the windings in the ignition coil overheated and shorted out. When the engine cooled down, the insulation restored itself and the coil performed as designed. I suspect both issues were in play at various times. Also, the single weight non-detergent oil, SAE 30, in the engine had a tendency to break down when overheated, and a rather thick layer of sludge formed in the engine oil pan, over time.

Various "remedies" had been tried over the years, such as higher flow water pumps to increase the flow of coolant through the engine and radiator, and a rebuild of the ignition coil to insert modern electronics within the old housing. Another trick was to add an electric fuel pump upstream of the mechanical fuel pump. The electric fuel pump "pushed" the fuel rather than "sucked" the fuel.

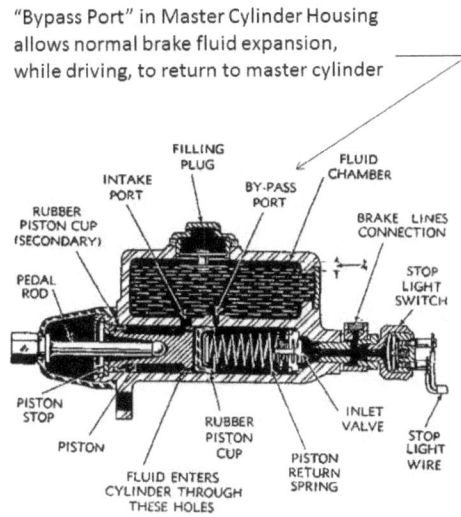

"Bypass Port" in Master Cylinder Housing allows normal brake fluid expansion, while driving, to return to master cylinder

Ref: Ford 1939 and 1940 Engine & Chassis Repair Manual

Over the years I owned the Ford, I added all these tricks. It didn't really help with the cooling. The one thing I didn't try was to retrofit an electric fan in front of the radiator. The fan operated either on a thermal switch or a manual switch. This retrofit is available.

Failure of the Ignition Switch

I started driving the Ford to meets with the North Shore Old Car Club (NSOCC) right after the restoration was complete. I was returning from one of those meets, on a weekday evening, near sundown, and was on a curving ramp off U.S. Route 1 leading to a back road. Suddenly the engine stopped.

I coasted to the edge of the curving ramp, but I was in a precarious position, as cars coming around the curve couldn't see me until the last minute. I wasn't able to do any troubleshooting, as I needed to stand in back of my car with my flashlight, waving cars around me. Suddenly, one of my car club members sped around the curve without stopping. I was furious for him not attempting to give me aid, as he was a good mechanic.

Finally, a car full of teenagers stopped and offered assistance. I had them wave the flashlight while, on a lark, I "hot wired" the ignition. It worked, and the car started right up. I drove home. The problem was with the ignition switch. It was a take-apart design, and the insulator was cracked, such it sometimes worked and sometimes shorted the ignition signal. I wrote a blistering letter to the NSOCC newsletter editor, and the club decided to equip future tours with two-way radios. I couldn't see how this would have avoided my situation, but I did all I was going to.

The Early Ford V-8 Club of America

I joined the *Early Ford V-8 Club of America* while I restored my 1939 Ford. The club had an outstanding magazine with many technical tips from owners of old Fords. Also, I was a member of the regional branch, the *Eastern Regional Branch of the Early Ford V-8 Club of America.* This club met in the basement of a church about a 50 minute drive from my area, making it difficult for me to attend monthly meetings. However, the club roster identified many tradesmen in the auto repair and body business. This was a resource I found valuable as my restoration progressed.

The Restoration

My car's restoration took seven years, mostly due to funding problems. I worked restoring small parts and cleaning/painting parts costing little money, in the early years. The work started in 1979 and finished in the spring of 1986. As it turns out, the *Early Ford V-8 Club* had its *Eastern National Regional Convention* in June of 1986. This was to be held at the Westborough Marriott just off the Massachusetts Turnpike, Interstate 90. Suddenly, I had a date goal to complete the restoration, in order to attend the convention with the freshly restored Ford.

The Eastern National Regional Convention of the Early Ford V-8 Club

I was determined to drive my newly restored car to the national regional convention. Two other local members were also going, so we drove convoy style, which gave me some additional confidence. Also, Marilyn followed the three old Fords in our modern Oldsmobile. The objective was to display my Ford, leave it in the hotel secure parking area and commute from home for the following few days.

The 40 mile highway drive to Westborough was uneventful, though I was nervous. Upon arrival, each car had to pass a safety inspection check before lining up in the display area by year and model. Unfortunately, my car was parked next to an identical model right down to the color of the paint. This "show car" was a professional restoration with no expense spared. It made my car look amateur. Part of the problem was my paint needed to be buffed-out to get the 20 mile-deep shine the lacquer paint was famous for back in the day. My shop tried to buff the paint but was unsuccessful. I learned it was necessary for the paint to cure for a few weeks prior to buffing. Naturally, the show car next to mine was buffed to perfection. As a means of comparison, I was into my car for about $25,000 by the time the restoration was complete. This comparison is made to the $40,000 restoration of the show car next to mine.

In the middle of the week, I decided to replenish the fuel in the gas tank. Upon return, I parked next to my clone as before. Later in the day, when I returned to my car, it was not where I left it. After a few minutes of panic, I learned the gas in my full tank had expanded with rising outdoor temperature, and was leaking out the overflow tube. The show car owner was nervous and enlisted the help of others to push my car to the far end of the parking lot, away from his baby.

My car is on the left and the one belonging to Jon Desmond on the right. Note the 1940 sealed beam headlights on Jon's car. Photo by Author

Late in the week, the team of trained judges performed their duties by examining every car in the competition. This team was as professional as could be, with every detail on the car evaluated against the standard of "as delivered." There were no points given for over-restoration. In some cases, over-restoration resulted in a deduction of points. These judges spent about 45 minutes to an hour going over every detail of each car. They even lay beneath the cars looking for discrepancies from "as delivered." While these inspections were taking place, the car owner typically paced back and forth nearby.

The Early Ford Club of America judging team awarded my car 899 points out of 999 total available. It was sufficient for a third place trophy. This was not bad considering the car was just out of the shop. I learned a few items on my car were incorrect vs. factory delivered, and I subsequently corrected them.

1939 Ford Convertibles lined up for inspection. Note the hood on Jon's car was not raised, as he didn't want his car judged for points. Photo by Author

Engine compartment of the 1939 Ford Convertible Coupe "show-car" parked next to mine. Judging by the pristine condition of the engine compartment, it was obvious this car was never driven over the road. It was driven onto an enclosed trailer, hauled to a car meet and driven off again to be displayed. The objective of owning an expensive car like this was to accumulate trophies. This was never my interest. Photo by Author

1939 Ford Convertible Coupe Deluxe Show Car. Photo by Author

In the foreground, notice a 1938 Ford Convertible Coupe. This was a driver equipped with dual exhaust and glass pack mufflers. It was a sweet running car. Photo by Author

Another view of Jon's car and mine. His car had dual exhaust, whereas mine had the stock single exhaust system. Photo by Author

Early Ford Woody Wagons lined up for display. Photo by Author

A pair of nice 1940 Ford Convertibles. Photo by Author

At the conclusion of the meet, we drove our old Fords home. It was dark, and my dash lights weren't hooked up yet. (They still aren't.) Marilyn had a flashlight, and periodically checked the engine temperature gauge as I drove. We pulled into a large parking lot to say our goodbyes, as we approached the point where we split up. As I alighted from my car, with the engine running, I noticed one headlight was not working. I learned later, the bulb was incorrect, and the tiny pip didn't make contact with the socket. It took a long time to fully debug the car.

Ford 1937-39 Passenger and Commercial Head Lamp Assembly

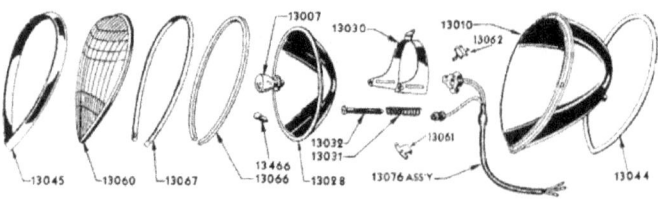

Note the take-apart headlight extant on the 1939 Ford. This was the last year, as 1940 Ford introduced the sealed beam headlight.

Tour to Nova Scotia

I was active in the NSOCC. A tour was planned to Nova Scotia, one summer. I was nervous about going so far in an antique car but agreed to go. The plan was to drive to South Portland, Maine, load the car onto the overnight ferry, The Scotia Prince, and board the ship. All went well and we had fun touring southern Nova Scotia. We made a trip to a private car collection in Plymouth, Nova Scotia and were returning to our hotel in Yarmouth.

I followed a friend in his Ford Model A. Suddenly, he made a panic stop on the soft shoulder of the highway, and I followed suit right behind him. As I came to a stop, my brakes didn't feel right. I asked him if everything was all right, and he replied someone else was following him back, and he lost sight of the other car in his rear view mirror. From that time on, my brakes were spongy. I observed a slight seepage on one of the front wheel brake cylinders, so I knew I had to be careful while using the brakes for the rest of the trip. I kept adding brake fluid to the master brake cylinder to keep the fluid level where it needed to be. We were in Nova Scotia the weekend JF Kennedy Jr. was killed while piloting a private plane to Hyannis, Massachusetts, along with his wife and relatives. We watched the news on TV in our hotel room.

Ford Model A Coupe
Internet photo Ref: http://en.wikipedia.org/wiki/File:1928-ford-archives.jpg

The Scotia Prince
Internet Photo Ref: http://en.wikipedia.org/wiki/File:Scotia_Prince_mg_5988.jpg

On the return trip, after we unloaded our cars from the ferry in South Portland, Maine, I asked one of my friends who had a 1941 Lincoln Convertible and who lived in the next town from me, to follow me home on the highway, in case I had a problem with the brakes. Except for a complete stop at a toll booth, where I had to use the hand brake, everything was uneventful. I was able to repair the brakes once I got home.

Event at the USS Constitution

The NSOCC had an event at the U.S.S. Constitution in Charlestown. It was a particularly hot Sunday. We were all huddled in the shade under a picnic area canopy, eating our lunch. The antique cars were parked in a row where the public could see them. Suddenly, a tour bus pulled up and discharged a group of Japanese tourists. Before I knew it, there was a tourist sitting in my car posing for another tourist taking his picture. I walked up to the one in the driver's seat, and asked him what he was doing sitting in my car. He was embarrassed; he thought the cars were part of a museum display and wanted a picture taken.

The author driving his 1939 Ford Convertible in a parade
Photo by Author

(Nightmare) Parade in Cambridge, Massachusetts

Another friend asked me to participate in a parade in Cambridge, Massachusetts, to commemorate Portuguese Day. I was to have magnetic signs on my two doors and Miss Portugal sitting on my rumble seat. This sounded like a good idea, and I agreed even though the location was somewhat out of my area. Everything went according to plan as we were directed to our location in the parade sequence. As the parade progressed, we had to stop at every cross street for an unknown reason, and wait until the movement reinitiated. I had mentioned previously the Ford flathead engine had a design defect resulting in a poor engine cooling situation, unless the car was moving at a clip of at least five to seven miles hour to aid in cooling.

The parade went on for two hours, and we made innumerable stops at cross streets. The engine temperature gage showed the cooling system was definitely upset with the slow progress. Finally, I knew the end of the parade was around the corner and down one more long street. We were barely moving, and I started to smell the sweet smell of radiator coolant as it went out the radiator overflow onto the street.

The temperature gage was all the way to "hot." I made a command decision, and pulled off to the side of the road, shut off the engine and raised the engine hood to aid in cooling the engine down. I turned to Miss Portugal and said, "Miss Portugal, you will have to get yourself a ride for the rest of the parade." Luckily, the modern car in front of mine had her parents as passengers, so she was able to make the transition easily.

As I was parked and my car was cooling off, a friend in his 1965 Mustang Convertible crawled by and suddenly threw up his engine coolant all over the engine and street. I guess I stopped just in time. This was the last parade I participated-in with my 1939 Ford. It was no fun when one had to closely watch the engine temperature gage constantly, while damning the slow progress of a parade.

Author and his wife, Marilyn, in period costume
Photo by Author

A Hooker and Her Pimp

I attended a cruise night at a local shopping center, in Danvers, Massachusetts when I spotted two unusual looking individuals walking across the parking lot from the shopping center to the lot where the collector cars were displayed.

Something didn't look right. The woman was Black, looked about

20, and was dressed to the hilt with a long dress and a slit all the way up to her hips. She was accompanied by a much older man, perhaps seventy years old, wearing shorts, white tee shirt and sneakers. The scene just didn't compute. When they entered the parking area for our cars, they began passing out business cards. She was a hooker and he was her pimp. This went on for a while, until the Danvers Police arrived. The two officers must have responded to someone's cell phone call. The police officers took the two entrepreneurs to the side and talked to them for a considerable amount of time. I actually lost interest. I think the officers tried to get the entrepreneurs to leave peacefully, and they were arguing. That was the first and last time I ever saw a hooker at a car show. Maybe I live a protected life.

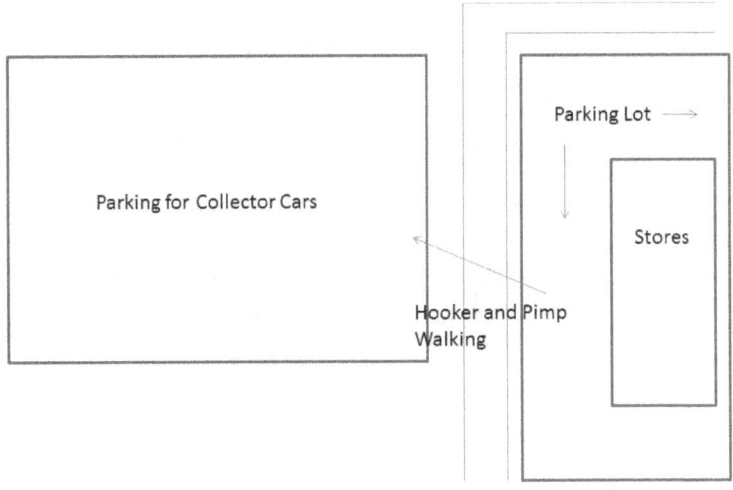

Diagram by Author

Year of Manufacture License Plate
Massachusetts passed a law allowing the "year of manufacture" license plate on an antique car. In other words, if I found an original 1939 license plate, I could register my 1939 Ford with this plate. There was one caveat. If someone had another year license plate

with the same number, the Registry would not allow the duplicate plate.

An acquaintance of mine in my club sold off his original antique license plates. He had a nice plate from 1939, and I bought it for $20. I had my insurance company make out the paperwork, and I went to the Registry. The clerk saw the plate and was all excited about the quality of the old plate. She called her supervisor over, and they both fawned over it for a while. Then the clerk looked up the number, and found another club member in Peabody had the same number, but in a later year. They wouldn't allow the registration.

Can't you hear the melodious exhaust sound from my 1939 Ford flathead engine exhausting through the Smithy Mufflers?
Photo by Author

Endicott Estates Massive Car Show

Annually, there is a large car show at the Endicott Estates in Dedham, Massachusetts. This is located on the south side of Boston. One year, I decided to attend. There was a heat wave, and I should have known better. The car performed satisfactorily on the run to Dedham. It was about 7:30 A.M. and the outside air temperature wasn't too bad yet. I got within a few blocks of the Endicott Estates and ended up in a massive backup of antique cars creeping toward the entrance. It took a while to get to the head of the line.

Midway, the engine temperature gauge showed full hot. I opened the hood, and crawled with the hood open. When I finally got to the entrance, I saw what the holdup was. The local auto club was collecting entrance fees, and everything was bogged down. I was so upset I ignored the money collector and drove right past to my parking spot. Nobody stopped me. Maybe they saw the steam coming out of my ears (and imminently from the radiator).

1939 Ford Dash
Photo by Author

By the time I left to go home, the outside air temperature was in excess of 90 degrees. While driving down Route 128, at relative high speed, the Ford periodically bucked as it was starving for fuel. The old vapor lock problem was present. I turned on the electric fuel pump to help push fuel to the carburetor, and managed to get home. That was the last time I attended this show. I understand the organizers of this show still have the problem of moving the cars through the entrance gate expeditiously.

Restoration Lesson Learned

The lesson I learned from this restoration experience was, it is cheaper to buy a car already restored than it is to actually restore one. On the good side, I learned every nut and bolt on the car during the process of research and restoration. Also, it costs about the same to restore a convertible, woodie (station wagon with wood sides) or coupe as it does to restore a four door sedan. However, the convertible, woodie and coupe finished product is much more valuable, due to the demand.

Movement of 1939 Ford to The Villages

When I moved to The Villages, Florida, as a "Snowbird," I shipped the 1939 Ford so I would have an antique car to enjoy during the winters. When I moved there full time, I shipped down my 1956 Oldsmobile Starfire. My middle child, Scott, had helped me complete the restoration of the 1939 Ford in the 1980s. Currently, my son and his wife also are living in The Villages. I gave my 1939 Ford to Scott as I have enough work to keep just one antique car running.

My Second 1950 Mercury

In 1980, while I was restoring my 1939 Ford Convertible Coupe, a process to take several years, I wanted an old car to drive around. I wanted another 1950 Mercury Coupe. I started searching and found an ad in the Boston Globe Newspaper for a car to fit my needs. The seller was Boston Mercury. I could not find Boston Mercury in the yellow pages under "Car Dealerships," but I called the telephone number anyway.

Boston Mercury actually was a person, who puttered around, buying and selling cars and parts, especially 1949-1951 Mercurys. I arranged to meet the owner at a storage building in Boston to look at the car he had advertised. The building was crammed full of antique cars stored by various owners in the Boston area. One of several cars he owned was another 1950 Mercury with license plates welded together to form a floor in front of the front seat, where the original floor had rotted out. I passed on that one.

He had another 1950 Mercury Coupe for sale meeting my needs. It was original, not restored, but was in remarkably good shape. It had dull chrome, some small rotted spots, but, overall, not bad shape for a "driver." I arranged to return the following week to give him time to move cars around in the storage building, and get the Mercury out and running. The next week, when I showed up and tried to start the car, the starter did not crank. Jon told me he put his "good" battery in the car, and I was to hold pressure with my finger on the starter button for several seconds, and the starter would eventually crank. I did and it did! The test drive was satisfactory. Jon wanted $2,500 for the car, but took my offer of $2,250. The condition of my purchase was he deliver the car to Peabody, where I lived, and I would give him a lift back to Boston.

1950 Mercury Coupe, purchased from John Desmond, *Boston Mercury*
Photo by Author

An interesting coincidence was he owned a 1939 Ford Convertible Coupe, the same car I was restoring. His car was a survivor, not restored, and was purely a driver. When I told him I bought the same car, he said he had looked at the car I bought, and passed. He didn't feel it was a good purchase, even as a parts car. He probably was correct.

I set out to make the Mercury more to my needs. I had a custom dual exhaust system installed, with glass pack mufflers. Boy did that engine sound nice! Next, I purchased a set of wide whitewall tires. The car radio didn't work, so I looked up an antique car radio repair shop in Marblehead. The owner told me there was a design defect in the 1950 radio model, and he would not recommend I repair my radio, but he had a restored 1951 radio for sale. The 1951 model looked the same externally. I paid $150 for the radio and put it in myself.

The directional signal lights did not work. Inspection revealed the wires were not hooked up to the switch under the steering wheel. I had a wiring diagram for the car, and knew from the color coding on the diagram, which wire went where. However, the color coding on the wires had faded, and I was not able to discern which color was which. I didn't hook up the directional lights.

My Ignition Coil Failed
I had attended a NSOCC meet at a member's house in Newberry, Massachusetts one Sunday. I just started out to drive home and the engine quit. Another club member came by and we tried to get it running, unsuccessfully. It was obviously an ignition problem. My friend went home, hooked up his car trailer to his SUV, returned and towed me home. Subsequent troubleshooting revealed the ignition coil had failed. I was able to buy a new one at an auto parts store and install it myself. While I was at it, I replaced all the ignition wires, spark plugs, distributer cap, rotor, points and condenser. I was not going to break down again.

My Car was a Movie Star
I was approached to have the car appear in a public television movie about to be shot, in Summerville, Massachusetts. I believe the name

of the movie was "The Education of Emily." The theme revolved around a man who decided to trade his Buick at a small used car lot, for a Hudson. The deal turned sour when the Hudson broke down as it was a "lemon."

Emily told her husband to return the car and get his money back. Her husband refused because there was no way a used car dealer would take back a car. Emily wouldn't take "No" for an answer, and successfully had the dealer reverse the trade. The movie was shot at a 1950s looking used car lot. My car was put on the lot in the front row as a prop. I had to be there for 12 hours while the movie scenes were shot. During that time, all kinds of unsavory characters walked by the shoot. I was paid "scale" which was $50 for my trouble. I used the money to buy a new six volt battery for the car. The movie writer was a former used car dealer. He disparagingly referred to my car as a "Junker." I didn't appreciate his comments, but I felt he was low class and not worthy of arguing with.

My Mercury and the Beach Boys
I took the car to work one nice summer day to show off to my coworkers. At lunchtime, I left the plant to get gas. As I left the gas station, I tuned in the local 50s radio station, and they were playing a "Beach Boys" song.

I was so pumped up I attempted to accelerate hard in first gear, drop the clutch and speed shift into second. As I did this, the shift lever suddenly went limp in my hands. I was disconnected from the transmission. Oh, Oh! I coasted to a stop on the side of the road. Across the street was Lynn Hospital. A gentleman saw me disabled and walked over. He was an old car buff. He put the transmission in second gear by moving the shift levers at the transmission, under the hood, and I nursed the car home in second gear about 10 miles. I located a replacement steering column with a good shift tube and replaced the unit on my car. Examination of the part off the car revealed the shift tube within the steering column had a spot welded "dog" and it had sheared off.

Transmission Lockup
Marilyn and I attended another club event, and were returning home in the Merc. I was within one half mile of the house, turned off the main road, and shifted from third gear back to second. I must have rushed the shift because the transmission couldn't make up its mind which gear to go into, and split the difference.

Basically, it locked up. A passerby, who was a mechanic, stopped and offered assistance. I jacked up the car by the rear end and we rocked the car back and forth until the transmission gears sought a neutral position. Unfortunately, the car fell off the jack, pried open the differential cover plate, and leaked the differential fluid out. I slowly drove the car the rest of the way home without fluid. I bought a used cover and serviced the oil, but was never able to completely seal this unit. I always had a small seepage.

I Sold the Merc
As my 1939 Ford was nearing completion, I sold the 1950 Mercury for $3,500. I broke even considering the initial cost of $2,250 and the cost of the improvements I made. The new owner painted the car, without fixing the corrosion, but it looked presentable. He also figured out how to wire the directional signals, so I have to give him credit.

I Thought the Mercury was out of my System
When I brought the 1939 Ford to the upholstery shop to have the upholstery and convertible top installed, there was a 1951 Mercury Convertible in the shop at the same time, having upholstery work done. I fell in love with the Mercurys of that era all over again!

Our 1956 Oldsmobile Starfire
I spent my first winter in retirement, December 2001 through March 2002, in a rented condo in Del Ray Beach, Florida. I always wanted a 1950's convertible. I surfed the internet, thoroughly bored, looking at what was for sale on eBay. Suddenly, I saw a listing and photographs for a 1956 Oldsmobile Starfire. It was the model identification given to the Oldsmobile 98 Convertible. I told my

wife, Marilyn, "Wow, I could be interested in that." She told me to go for it. She didn't have to tell me twice.

I Purchased the Car

I found an email address in the eBay listing, and contacted the owner. He and the car were located less than four miles from where I was sitting. I was in Del Ray Beach, Florida and the car was in Boca Raton. I made arrangements to look at the car, while Marilyn sat in our Honda.

It was immediately evident there was a fuel delivery problem, as the engine needed to be cranked excessively before it fired. Then, the engine didn't take the fuel readily under acceleration. Also, the transmission shifted hard going into 3rd gear.

As the body, upholstery, and paint were in excellent condition, I decided to bid on the car. This was the first time I ever bid on anything on eBay, so I was uncertain of the procedure. I waited until the last five minutes, and put my bid in. I didn't appreciate I needed to submit a credit card number to validate my bid, because of the high value, and it ate up precious minutes. I was not sure the process would validate me in time, but it did. My bid broke through the reserve, I was high bidder, and I bought the car. The purchase price was $30,000. I was determined not to pay for the car out of family funds. I went back to work at GE, as a retiree callback, in my former specialty, during the summer, to earn the money to pay for the car. It took one whole summer and half the next to finish paying for the car.

Statistics:

Nickname: Big Bertha (Marilyn and I named her.)

Engine: 324 Cubic Inch Overhead Valve V-8, 240 HP with 4 Barrel Rochester Carburetor; "Oldsmobile Rocket Engine"

Transmission: Jetaway 4 speed Automatic

(This was the improved version of Hydramatic introduced for 1956 Olds 98 and Super 88)

Weight: 4325 lbs.

Cost when new: $3740

Quantity Built: 8531

Power Assists: Windows,   Brakes,   Front   Seat,   Radio   Antenna,
Convertible Top

Gas Requirement: High Test

A friend commented. "The good news is this car has all power accessories. The bad news is
they are all first generation."

Getting the Car in Roadworthy Condition

I contracted with a trucker to transport the car to my home in
Massachusetts.   This trucker had a fleet of specially designed
enclosed auto transporters to move antique, classic and exotic cars.
The cost was approximately $1100. When I returned home, I nursed
the car over to a friend, a retired auto mechanic, to straighten out the
problems.

The rebuilt carburetor had a defective (but new) accelerator pump
umbrella seal. When the throttle was pumped to deliver a squirt of
fuel, the "umbrella" collapsed rather than pushing the fuel, and the
fuel simply cycled around the seal.

The vacuum advance diaphragm was compromised (Rubber
diaphragm had failed).

The mechanical fuel pump was inoperative because the lever was too
short. It looked like it was sheared off and it definitely was not being
activated by the camshaft.   I discovered this condition when I
purchased a rebuilt mechanical fuel pump, and I removed the old one.
The car also had an electric fuel pump, as an accessory, and it was
the primary and only operating pump. I pray the missing length of
pump lever is not on the bottom of the oil pan.

The transmission was another story. I had a trusted transmission
shop tear it down and rebuild it. The transmission was the Jetaway

Transmission. It was new for 1956, and was a variation of the Oldsmobile Hydramatic. It was standard on the Oldsmobile 98 and optional on the Oldsmobile Super 88.

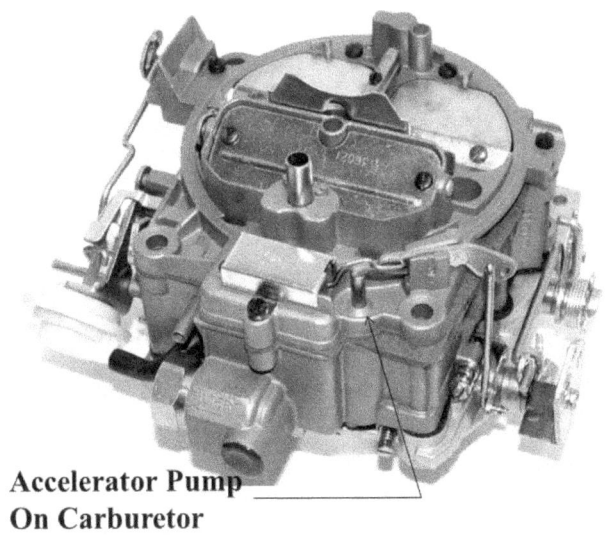

**Accelerator Pump
On Carburetor**

Rochester Quadrajet Carburetor. Ref: en.wikipedia.org

Our 1956 Oldsmobile Starfire; Color is Two Tone Bold Red and White
Photo by Author

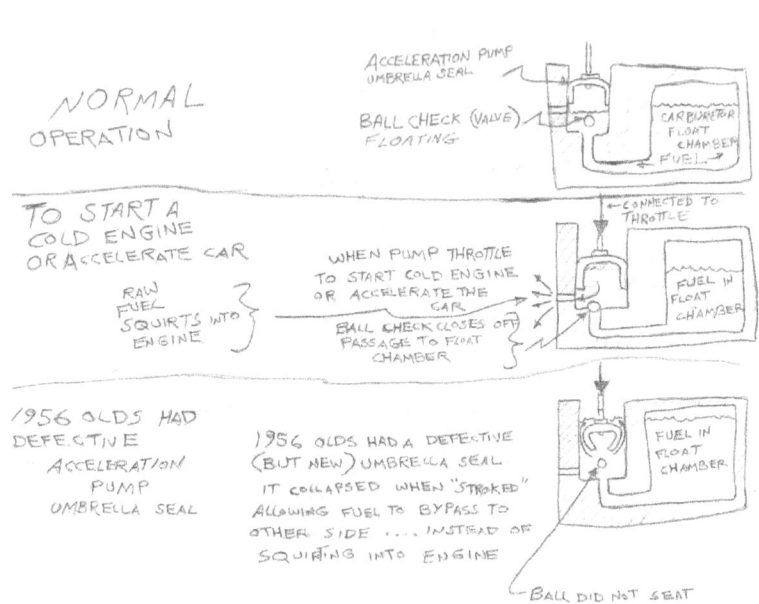

Defective Carburetor Acceleration Pump Umbrella Seal
Sketches by Author

It relied on a second torque converter to fill or drain the transmission fluid to shift from 1st to 2nd and from 3rd to 4th. Second to third relied on the conventional band, and it was the hard shift. It took the whole summer to straighten the Jetaway out, including getting another transmission case, but when it was done, it was right. It cost me just under $2000.

The Addition of "Super" to the Transmission
The Oldsmobile Super 88 and 98 (including the Starfire, which was the Oldsmobile 98 convertible) had what appeared to be an extra gear in the automatic transmission. On the transmission selector, the option "S" or Super was added to P, N, Dr, Lo and R.

The Oldsmobile marketing department advertised this as an

additional gear called "Super." When the driver wanted to strongly accelerate, the gear shift lever was positioned to "S" and the car shifted into "Super" allowing the car to strongly accelerate. It was all marketing hype. Positioning the gearshift to "S" simply dropped the transmission gear to third gear and prevented it from shifting back to fourth, thereby allowing the engine to accelerate the car faster than in the higher gears. There was no extra "Super" gear!

The Restoration Shop
The car had been "professionally" restored on speculation. It was disassembled, had some metal restoration work done, and was painted. It then sat around in the shop disassembled for about two years. They must have rushed the reassembly, because they took too many shortcuts. One was the reuse of original rubber hoses on the chassis, when new and inexpensive reproduction replacements were readily available. Specifically, the transmission oil cooler hoses, the hose from the master brake cylinder to the tee, power steering pump hoses, and so forth. It would have been easy and inexpensive to replace these parts when the body was off the frame. However, replacement of the parts with the body on the frame was something else altogether.

I assume the car hung around the shop for a couple of years, disassembled. Then, someone of authority must have said, "Get that !&** thing out of here," and assigned the person with the lowest hourly pay rate to put it together. The seller in Florida had business contacts with a major auto manufacturer, and I know he tried to have a local dealership straighten out the problems with the car. However, modern mechanics are not familiar with old cars, and the repairs didn't work out. It took many trips to my friend, a retired auto mechanic, to correct these errors.

In 1956, Artwork was used in promotional literature for the new cars. Later, photographs were used.

Oldsmobile Promotional Literature for 1956

"Super" Gear

Problems Exhibited by our 1956 Oldsmobile Starfire and Solutions

New Convertible Top Adventure

The convertible top header didn't fit properly at the right front interface with the windshield. Whoever restored the convertible top cut the material too short, and the header could not be adjusted to fit properly. I bit the bullet and scheduled the car into a highly recommended shop to have the top replaced. The cost estimate was $1000.

A week later, I received a cell phone call, on Sunday, from the owner's sister who worked in the office. There had been a spontaneous combustion fire in the shop over the weekend, and there was soot and smoke damage to all the cars in the shop. It took a year, but when the insurance company straightened out the dollars, and the shop stripped and rebuilt the car, it was right as it could be. The car had originally been restored with red vinyl upholstery. The convertible shop owner felt so bad he reinstalled the upholstery with red leather, the way the factory had originally built the car. The insurance covered all new carpeting, door panels, padded dash, trunk carpet, and so forth.

The insurance settlement was $9200. I sat down with the shop owner and we compared notes as to what he had into the car vs. the insurance check. He discounted his investment to agree with the insurance.

Radial Tires Were Dry Rotted

I got the car ready for an end-of-the-year car show in New Hampshire. While I cleaned the radial whitewalls, I noticed some fine cracks oriented radially around one tire. Close inspection revealed all four tires had the cracks. I called the tire manufacturer, and explained I was not the original owner, but the tires had probably less than 1000 miles on them, which was the truth. They stepped right up and sent me four new radial whitewalls at no charge. My cost was to replace the tires on their rims, and ship the defective tires back to the factory. I felt I was treated fairly by the tire manufacturer, and would highly recommend them.

Author's 1956 Oldsmobile Starfire after the Smoky Fire
Photo by Author

A Restored Car is not always <u>Completely</u> Restored

One thing I learned about buying a restored car. It is never really 100% restored unless you do the restoration yourself. I spent the next five years pecking away at all the errors of omission and screw-ups the restoration shop had made. I replaced everything from the windshield wiper motor, convertible top electro-hydraulic pump and actuators, power seat motor, and so forth. A restoration shop usually replaced only the items they felt needed to be replaced. The rest of the original parts were just waiting to fail.

Chronic Problem with Brakes

I was concerned about the power brakes for several years. One year, Marilyn and I took the car on a tour to Nova Scotia with the NSOCC. We drove to South Portland, Maine, where we loaded the car onto the overnight ferry, The Scotia Prince, and boarded the ferry. The trip was enjoyable until, several days later, I followed a 1960 Chevrolet Convertible down a long hill terminating at a stop light, in Halifax, Nova Scotia.

After riding the brakes for a while, to maintain my distance, I attempted to actually stop the car before I hit the Chevy at the light. It was then I experienced complete brake fade. The car had a nice hard pedal but no braking action. I stopped just barely before I hit the Chevy. After the brakes cooled down, they behaved normally. When I returned home, back I went to my friend, the mechanic. When the rear wheels were pulled, he noted the rear brake shoes were soaked with differential fluid. Both rear wheel seals were failed, and I was braking using only the front brakes. I had been driving the car this way ever since I bought it, several years earlier. Replacement of the rear wheel seals and cleaning of the rear brake shoes put me back in business.

I still was not happy with the braking action, and several seasons later, the pedal was now almost to the floor. I never was happy getting under this heavy car, either on jack stands or on homemade ramps, so I decided to seek professional help.

The problem was with the right rear steel brake line. The line obviously was new, but it was more or less a straight shot, and was just long enough to make the connection, but not long enough to actually seal with the mating part. It had been weeping for years, until too much air got in the system. Replacement of the line and a good bleeding of the system gave me the pedal hardness I never had since I bought the car. Finally, the power brakes worked the way I remembered from the 1950s.

The Car Leaked Water
Upon returning from Nova Scotia, the club had reserved a group of rooms at a South Portland motel, as the arrival time was at dusk, and most of us didn't want to drive the old cars home in the dark. The next morning, it started to rain as we departed for home. The trip

was mostly south on Interstate 95, so we were going at highway speed. The rain got steadily worse. This was when I discovered the windshield wipers did not work. By the time we got home, the carpet and trunk matt were thoroughly soaked. This was how I discovered the car leaked. The restoration shop did a great job (not), and I would not recommend them.

10% Ethanol in Gasoline

Along came 10% Ethanol in the gas. Modern cars had no problem with the Ethanol, as they used rubber parts in their fuel systems compatible with the fuel. Also, the computer controlled fuel injection handled the Ethanol. But for the old cars, the fuel stayed fresh for as little as 30 days before the alcohol started to come out of solution and attract water. Old cars not used regularly started having problems with rust and debris in the fuel tank. Also, Ethanol deteriorated the rubber parts in the old car's fuel system. Suddenly, most old cars were having major problems with the gas. Additives such as Stabil are available to help, but old cars continue to have problems.

My first experience with the quality of the gas came when I took the car out of winter storage, one year. After charging the battery, I started the engine. It ran fine for a couple of minutes, but then ran erratically. I checked everything in the ignition system, and everything checked out. As I was out of ideas, I decided to pump the gas tank nearly empty, using my electric fuel pump and three gas cans I purchased for this purpose. With a nearly empty gas tank, my son drove the car to a nearby gas station and filled the tank with fresh high test gas. As soon as he pumped fresh gas and drove the car away from the gas station the engine smoothed out and ran as originally advertised.

1956 Oldsmobile Dash
Photo by Author

Engine Cleanout Procedure

A couple of years ago, once again I complained about a poorly running engine. Back to my friend I went. He put the engine on his dynamometer, and told me I had several valves not working properly. I saw dollar bills floating out of my wallet. My friend told me of an old timer's trick.

He had me purchase a quart of two cycle engine oil and mix it with five gallons of gas. Then he had me put this combination in the gas tank and fill the rest of the tank with gas. I was to take the car on a highway trip to burn up the mixture in the gas tank. Since I was going to a car event in New Hampshire that weekend, I followed his recommendation. Immediately, the engine just didn't run correctly on the gas mixture. The engine ran even worse when I traveled on the back roads in New Hampshire, and encountered stops at traffic lights.

I nursed the car back home and continued to burn most of the gas in the tank. I thought when I added fresh gas to the nearly empty gas tank, the engine would suddenly run great. It didn't happen. The engine continued to run poorly. My mechanic was consulted by telephone. He told me to turn the two carburetor mixture screws all the way in until they bottomed, while counting the number of turns. Then, remove the screws altogether, and use my compressor and air gun to blow out any debris in the carburetor through the two holes where the mixture screws were. Then put the screws back to the same number of turns I had counted originally. As soon as I started the engine, the improvement in engine performance was obvious.

Nightmare Tour to a Fifties Diner in New Hampshire
The NSOCC scheduled a trip to a 1950s diner in Rochester, New Hampshire. It was a beautiful sunny Sunday, and I did not check the weather forecast. The group met at a rest stop on Interstate 95 North near the New Hampshire border. We departed convoy style, and the further we drove into New Hampshire the darker the sky appeared in the distance in the west. Soon, there was lightning in the distance. Then the storm was upon us with wind gusts and heavy rain. We saw tree limbs on the ground as we continued, but we made it to the diner.

After lunch, we were given directions back to a major highway, and set out individually to go home. We didn't get more than a half mile before we encountered fallen trees and a detour sign. We were detoured into a residential neighborhood and then the detour signs disappeared. We were on our own. After becoming thoroughly lost we saw one car from our group, a 1972 Ford, going in the opposite direction. I decided the driver of the Ford knew where he was going, and reversed course and followed him. Then it started raining again, and it got worse. My windows fogged up and visibility was poor.

It seemed like hours, heading west down back roads, looking for any sign to a major north-south highway. Finally, I spotted a fire station and pulled into the parking lot. The station was open and unoccupied, but there was a map on the wall, and I saw U.S. Route 28. It was just one more block. To make a long story short, I burned most of a whole tank of gas that day, and once again soaked the car rugs and trunk with rain. By the time I arrived home I was upset. It was my fault for not checking the weather forecast. This was a valuable lesson, and I now religiously check the weather before taking out Big Bertha.

Car Replacement Cost Appraisal

Before I moved to The Villages, I had the Oldsmobile professionally appraised. I raised the insured value to agree with the appraisal of $75,000. Ten years ago I purchased the car on eBay for $30,000. This was not a bad investment, considering the fun we had with the car over the 17 years we have owned it.

The Moral of the Story

What is the moral of the story? There is no moral. I am in a hobby more obsession than hobby. We car nuts buy what we remember, and we do what we can to keep the car running, and enjoy it as much as possible. Oh yes, this car was manufactured the year I graduated high school. I drove the car to my 50th high school reunion, Class of 1956! I was the hit of the reunion.

Chapter 13; REFLECTION ON MY CAREERS

Air Force Career

I consider my Air Force career as the high point of my life, professionally. It was a complete surprise to me I qualified to go to Undergraduate Navigator School, as I assumed I was disqualified because I wore glasses. I consider myself lucky to be born in just the right time frame, such as I was able to be a navigator before the need was superseded by the satellite enabled GPS (Ground Positioning System) which is available to everyone now. The jet powered KC-135 tanker aircraft, equipped with navigators when I was on active duty, have been retrofitted with the electronics to eliminate the manned navigator. And, yes, these aircraft are still flying.

The years I flew on active duty, in SAC KC-97G Tankers, and part time, in the Air Force Reserve in the C-124C and in the Air National Guard in the KC-97L were the best years of my professional life. On active duty, I was impressed with the dead seriousness of being a part of the massive retaliatory bomber/tanker force in SAC.

Later, what I loved the most about the Air Force Reserve C-124 was the unit's mission. We were tasked to fly outsized cargo for the active Air Force all over the world, in an aircraft that flew at 200 Knots True Airspeed, unpressurized, at 9000 or 10,000 ft. altitude. Also, the aircraft autopilot incorporated an altitude hold feature, which made for a more stable aircraft from which to sight the sextant for celestial navigation. What I didn't like about the C-124 was the worn out condition of the aircraft, and the poor aircraft performance at the weights we were tasked to fly them.

What I loved about the Ohio Air National Guard's KC-97L Tankers was the improved aircraft performance with the addition of the General Electric J-47 engines under the wings on pylons providing additional thrust. As I stated previously in the book, the takeoff aircraft performance was spectacular, what with four powerful 3500 horsepower piston engines and two jet engines combined with a more reasonable takeoff weight, compared to the requirements of SAC on active duty. The Air National Guard also added altitude-hold to the autopilot; the SAC aircraft did not have this feature. The Creek Party missions to Germany were a challenge for the navigator, who needed all his skills during the over water portions of the flights. Over the years, I accumulated 1500 hours flying time in the KC-97 aircraft, total, both on SAC and Air National Guard aircraft.

I went to an open house for the Coast Guard unit at the former Barbers Point Naval Air Station, when I was in Hawaii. They had a C-130 aircraft on static display. I asked the pilot, "where is the sextant?" He didn't know what I was talking about. I felt old!

Peer Competition in the Reserve Forces
A rated navigator flying in the Air Force Reserve or Air National Guard stayed on a crew for his/her whole career, with minor exceptions. I found this much more to my satisfaction, and this fact did not hamper my chances for promotion, as it would have while on active duty. I still had to take the professional military education courses, but there was no stigma to taking the courses by correspondence, which I did. I completed Squadron Officers School and Air Command and Staff College by correspondence. I drew the line with Air War College, as this course required independent study and a thesis, which I did not have the time or inclination for, because I was so involved with my engineering career at this point. The latter course was a requirement for promotion to full Colonel, and I made the conscious decision I would not compete for this promotion. Basically, there was no active peer competition while flying in the reserve forces.

Air Force Responsibility for a Young Rated Officer
I found it amazing the level of responsibility bestowed upon very young rated officers, both pilots and navigators. They had virtually life or death responsibility piloting and navigating aircraft at weights not envisioned by the aircraft manufacturers over long over-water expanses. Once the navigation radios broke lock (about 100 nautical miles from the tip of land location, during coast out), the navigator was all by himself. He either did his job as trained, or he screwed up,

all on his own. The emotional high, when the radar picked up the landmass across the ocean, on the other side, from about 150 nautical miles range, and your aircraft was very close to being on course, was indescribable. The pilots and navigators in the bomber aircraft had the additional responsibility for nuclear weapons! I never had anywhere near this level of responsibility as an engineer working in industry.

Engineering Career

My position at Hamilton Standard, as a propeller installation engineer, exposed me to upper management, as well as provided me with interesting assignments. I covered the first flight of the CC-109 twin engine Cosmo at Pacific Airmotive in California, and was the lead engineer on-site for both prototype C-130A conversions, both applications powered by the 54H60 propeller system.

I expected the same exposure when I left Hamilton Standard to work for GE. Unfortunately, GE operated differently. There was no way a line engineer from GE would be trusted to cover the first flight of a new application; that was entrusted to upper level "Project Engineers." It became obvious to me early at GE I was to be just another number in a large pool of engineers. I spent the rest of my GE career, of 31 years, looking for that brass ring of exposure, and never found it.

My biggest mistake while working at GE was not taking advantage of an opportunity offered to me. I worked my way up to senior engineer in external configurations design, and looked to worm my way out of design engineering into a management role. I had unsuccessfully competed for an open posted job in the J85 Engine Project organization. This organization was the prime engineering interface between GE and the engine customers, both airframe manufacturers and end use customers, and had high visibility to upper management. Project jobs were coveted and were considered upper level positions. I requested feedback as to why I was not successful. My engineering manager took it upon himself to meet with the J85 Project Manager to find out why I did not get the position. He met with me later and told me the primary reason was I did not have engine systems and performance experience.

He then made me an offer I could not refuse (even though I did refuse). He offered to move me into the J85 Engine Systems and Performance Organization, under him, to get the requisite experience. He told me it would not make any difference in my pay. As I said earlier, I had just attained the senior engineer rating in my primary specialty, which entitled me to sit in my own cubicle instead of in the engineering bullpen, and I didn't want to give that up. I politely refused the generous offer. I lived to regret this decision down the road!

My experience working at GE was it was easy to become pigeonholed in an engineering specialty. I will use GE Aircraft Engines as an example. When hired, the engineer was assigned to a particular design engineering group.

Some typical design engineering groups at GE Aircraft Engines were as follows:

Structures, Cold, such as fan frame, front frame, etc.

Structures, Hot, such as combustion casing, turbine frame, etc.

Bearings, Seals and Drives, such as gearbox, bearings, etc.

Engine Controls

External Configurations Design, such as external piping and brackets

Company Career Advancement Hold Back
It is to the company's advantage to keep the engineer in a particular design group, as the cross training costs are reduced and the need to have a trained engineering work force stabilized. However, it is not to the engineer's advantage to remain in one group. He/she will progress in experience and will eventually rise to senior engineer (expert in the specialty) or senior staff engineer (program manager), and then the engineer's career stagnates. Mid-career the raises

stagnate and salary raise intervals become stretched out. Responsibility increases greatly with these titles, but compensation does not. Later, salary raises become fewer and at longer intervals still. The company knows a high service employee is not going to leave for another company, and therefore, the company allocates the raise dollars to the younger engineers, who are company mobile.

At GE, the senior engineers (I was a senior engineer earlier in my career) in a particular specialty have the backbone technical expertise to insure technical excellence in the final product. The senior staff engineers (I was a senior staff engineer in my later years) have total responsibility for the engineering program from cost analysis, engineering guidance for the younger engineers under him/her, drafting coordination and manpower implementation, management reviews, and technical implementation through final product qualification. In each specialty, these senior staff engineers provided services as if they ran their own businesses; they recruited design engineers, as necessary from within, negotiated drafting support, provided upper management with progress reports, and so forth. I cannot complain about the salary I was paid for my services, as a senior staff engineer, as I led an upper middle class life. However, the level of compensation <u>commensurate with the value to the company</u> was unfair!

The Struggle for Upward Mobility
The day-to-day head butting occurring in a large organization such as GE top heavy with talent, all vying for recognition and promotion into limited available openings, made one sick. It was important to be seen. An example of this was a female engineer who the company needed to move up. (affirmative action or the newer buzz word, diversity). She sat up front at every meeting she went to, and made sure she asked a question at the end of the meeting, regardless of the need for information; the objective was to be noticed. It apparently worked, as she was promoted into a group in which she had no experience.

While I am on the subject, I worked in the external configurations design group for much of my career at GE. This was a group not noted for promoting from within. Whenever there was an opening, an outsider was backfilled. Then management complained engineers tried to move out of the group constantly. This was especially true of the younger engineers who saw the handwriting on the wall concerning lack of promotion from within. It was difficult to understand what was so very obvious to the rank and file was so impossible to see from upper management.

One way engineers felt they could short circuit the upward mobility barrier was to leave the company and go to work for a competitor, taking with them their experience built-up over years. I have felt this works for a short period of time, and the new company takes into consideration the step increase when the next raise is scheduled. One thing to be especially wary of is to question why nobody has applied for a position from within the new company, especially in a new product line, which usually is a coveted position. Many times the reason is the hiring manager is a problem personality and everyone within knows it. I personally observed this situation at GE.

There is one other consideration when changing companies. They usually hire engineers during boom times. My experience told me whenever times are really good, in the aerospace industry, a crash is just around the corner. And when that crash occurs, the engineers laid off are usually the ones with the least company service.

Engineering Arrogance

My observation of engineering organizations over many years drove me to the conclusion middle and upper management personnel (former engineers) had a unique quality; arrogance! Time and time again I observed this characteristic.

I lead the technical effort to redesign the F404 afterburning engine's external configuration to power the Korean T-50 Golden Eagle Pilot Trainer and Lead-in Combat Fighter. Almost the entire package of external piping was redesigned for this single engine application. Both the schedule and the technical aspects of the program were challenging. Our department general manager (one step below vice-president) made a visit to Lynn from *heaven* (Evendale, Ohio), and all the senior staff engineers briefed him on the status of their programs. I made a comprehensive presentation on the technical challenges as well as the tight schedule, and summed it up by saying, "Hopefully, we will meet the schedule." This pompous ass responded with a raised eyebrow, "Hopefully?" I was so bullshit I couldn't see straight. He wouldn't raise one finger to help me bring in the development hardware on time, and he was busting my stones on a phrase I used. What audacity!

Another example of this occurred during one of my work sessions as a retiree callback during one summer. I worked on a vibration problem exhibited on an installed F-404-GE-102 engine in the Korean T-50 Pilot Trainer (I was the senior staff engineer on this program when I was still working full time, for external configurations design.) I attended a meeting where the systems manager directed engineering to conduct a vibration shake table test of a proposed fix to demonstrate infinite life (10 to the sixth power vibration cycles). Engineering spent considerable dollars designing and building a heavy metal fixture to support the redesigned tube assembly on the shake table. Weeks later, when engineering reported on the progress of the fixture, the manager who directed the effort dumped all over engineering for spending the money on a fixture for a test that he did not condone. Unbelievable!

During my active work years at GE, there was a section manager (engineering manager of a product line) who epitomized the height of arrogance. The company had already instituted a ban on smoking in the workplace, but this manager didn't think it applied to him. He chain smoked in his office, in the conference rooms, etc. Nobody called him on this. This manager had no use for anyone who didn't have immediate answers to his questions. He held daily briefings on the status of various aspects of the designs, and fed rapid-fire questions to the briefer. If the poor subordinate didn't know the answer immediately, he threw him out of his office using the "F" word liberally. He ran a terror operation.

This manager mandated everyone work a minimum of four hours overtime a week, for no pay. Other engineering groups were also working four hours overtime per week, but the engineers were paid (only straight time earnings for the overtime). I was vocal about never working for this individual, and, luckily, I never had to. The manufacturing engineers who were responsible for bringing in the hardware to support an ambitious engine test program, worked directly for this manager. The tension was so great one manufacturing engineer had a massive stroke while working on this program. Another suddenly stood up on his desk and started to do a strange dance. This program was successful because it was run on fear alone.

Another example of arrogance is cited. Once Lynn Engineering was made a satellite of Evendale, Lynn's ideas were worthless. During design reviews with the chief engineer's office (in Evendale, of course), if a design concept was not proven on an Evendale engine, it was no good. I designed a flexible joint in a short stubby fuel tube. During the design review, the reviewer stated I should use the particular flexible joint used on the Evendale F110 engine (on the F-16 fighter aircraft). Subsequent research indicated the similarly was nonexistent; the F110 engine used a completely different approach that didn't apply, but if it wasn't used on "their" engine, it wasn't any good! What arrogance!

High Potential List

GE Aircraft Engines had a _**list**_ of personnel who were considered _**high potential**_. The company denied this list existed, but I learned from an unimpeachable first-hand source this list did, indeed, exist, and it was kept under lock and key. People on this list did not compete with peons for open jobs in the open job posting system. I am convinced the candidate who got the job as a propeller installation engineer for the CT7 engine, working with Hamilton Standard, and who aced me out of this job even though he had no experience, was on this coveted list. He was put in this job because it had high management visibility, and he could get the experience by working in this field.

Engineering Compensation

When I was a young engineer, I was of the impression aerospace engineers were compensated well because their profession required exceptional dedication to excellence, rapid response to problems with the hardware they designed, and the dire consequences of failure of their hardware in aircraft and spacecraft. I felt this was true in the early part of my career with GE. Engineers in the jet engine division were compensated about one level higher than engineers in the appliance division, for instance.

When Jack Welch took over the company, that distinction was eliminated. Jet Engine Division engineers were eventually downgraded, using the salary compensation system, one level, to make all engineers on a level playing field. It was about this time I became aware engineers were not compensated all that well compared to other professions.

The following is written to demonstrate the level of responsibility given to engineers. The engineer applies his/her expertise to design hardware so the company has a product to sell. ***Without the engineer, the company does not have a product to sell***.

This design process starts with an engineering concept, followed by analysis, and gets fleshed out in an engineering layout, development hardware is procured, production drawings are generated, component tests are accomplished, if necessary, followed by full scale engine tests. Appropriate design reviews are held with the chief engineer's office. Engineering documentation is then accomplished followed by submittal to the governing authority for acceptance, such as the FAA or military.

There is a lot more to the design process, but I abbreviated it for brevity. In exchange for that contribution, the engineer is treated as hired help and paid a salary just high enough to prevent him from leaving, until the engineer gets fairly high in seniority, when the company can screw him over with impunity.

Engineer vs. Drafting Compensation

Engineer, Set Refueling Power

In GE Lynn, the drafting organization was unionized. Accordingly, only draftsmen could make an engineering layout of a new design. Engineers could make sketches of what they wanted the design to look like, but could not use drafting tools to make straight lines, French curves to draw curves, and other view projections. The way it worked was, the engineer requested a designer (draftsman) be assigned. Once that happened, the engineer worked side by side with him/her, almost holding his/her hand as the design progressed.

The union mentality in drafting, in my later years, was the same as for any union shop; keep productivity low, maximize days off, even counting sick days as additional vacation days, and drag the feet to require overtime for job completion. As the older senior designers retired, the new crop never had the interest or talent of the older ones. When I was still in design engineering, I would have to hand hold the designer, as to what I wanted on the design layout. I could not expect to provide minimal direction, and have the output as I expected. The fact of life of senior designers (draftspersons) being paid nearly what I made stuck in my throat! And, as with any union worker, the longer the engineer sat with the draftsman, the slower the pace of the work

Human Resources (Personnel)

I have nothing but contempt for every Human Resources representative I have ever been exposed to at GE Aircraft Engines, and I have had dealings with many. It seems their function is to look out for upper management and keep the aspirations for engineers on the firing line low. I will quote an example. I broke out of design engineering into program control (for a brief period of time) after eight years in external configurations design. My manager, in design, told me he had forgotten to sign-off on a salary increase I was scheduled for. The significance was I left design, and expected the salary increase to show up in my next paycheck, while working in program control. When the increase did not appear, my former manager realized his error.

I found out this situation was covered by company policy; all I had to do was get a letter signed by my new section manager, which I did. My human resources representative telephoned me and respectfully requested I withdraw my letter. Her boss forced her to do this, as it would set a **bad** precedent. This would have cost the company just two weeks of the actual increase in salary. I think GE could afford it, and it would have been the right thing to do.

Under layoff conditions, human resources representatives tell engineers to go find a new job. They put the onus on the engineer to scramble to find another position. I have always felt finding a new job was in their job description. This was not based on one example; I observed this attitude time and time again.

I will admit, if an engineer could not find a job, and was in danger of going out the door, sometimes the human resources representative might step in at the last moment and provide minimal guidance as to where a job might be lurking. However, when a whole engineering project is cancelled, human resources will go out of their way to find new positions for upper level management.

When I was having trouble finding a position outside of engineering, human resources told me I was not "flashy" enough. Apparently, being "flashy" was a requirement for moving up. Silly me, I thought hard work and facilitating results was important, but I was wrong; it was more important to be politically connected and be **SEEN** than it was to put out the hard core work. WOW!

Aerospace Engineer Responsibility
I described the responsibilities bestowed upon Air Force rated pilots and navigators, early in their careers. In engineering, this was not the case. Young engineers were taught which equations to use in every situation. The engineering best practices manual had to be followed, and appropriate engineering reviews needed to be accomplished with the chief engineer's staff at appropriate intervals.

In other words, the engineer was never free to innovate. I had to laugh, on the engineering checklist filled out for every engineering change application, one item requested was if there were any ideas worthy of patent application that came out of the change. How could there be a patent if the engineer was not allowed to innovate?

I had scheduled a design review of an external configurations design with the chief engineers' office, at the full scale mockup shop. When the chief engineer's representative showed up, one of the mockup shop foreman said, "Here come the *why did'nt yers!*" In other words, "Why didn't you do this, why didn't you do that?" That about summed it up concerning what a design review is all about.

Engineer, Set Refueling Power

www.ingramcontent.com/pod-product-compliance
Lightning Source LLC
Chambersburg PA
CBHW071247220526
45468CB00001B/30